Pier Scour in Clear-Water Conditions with Non-Uniform Bed Materials

FOREWORD

The study described in this report was conducted at the Federal Highway Administration's Turner-Fairbank Highway Research Center (TFHRC) J. Sterling Jones Hydraulics Laboratory in response to State transportation departments' requests for new design guidance to predict bridge pier scour for coarse bed material. The study included experiments at the TFHRC J. Sterling Jones Hydraulics Laboratory and analysis of data from the Colorado State University and the United States Geological Survey. This report will be of interest to hydraulic engineers and bridge engineers involved in bridge foundation design. It is being distributed as an electronic document through the TFHRC Web site (http://www.fhwa.dot.gov/research/)

Jorge E. Pagan-Ortiz
Director, Office of Infrastructure
Research and Development

TECHNICAL REPORT DOCUMENTATION PAGE

1. Report No. FHWA-HRT-12-022	2. Government Accession No.	3. Recipient's Catalog No.
4. Title and Subtitle Pier Scour in Clear-Water Conditions with Non-Uniform Bed Materials		5. Report Date May 2012
		6. Performing Organization Code
7. Author(s) Junke Guo, Oscar Suaznabar, Haoyin Shan, and Jerry Shen		8. Performing Organization Report No.
9. Performing Organization Name and Address Genex Systems, LLC 2 Eaton Street, Suite 603 Hampton, VA 23669 Department of Civil Engineering University of Nebraska-Lincoln 1110 67th Street, 200E Omaha, NE 68182-0178		10. Work Unit No. (TRAIS)
		11. Contract or Grant No. DTFH61-11-D-00010
12. Sponsoring Agency Name and Address Office of Infrastructure Research and Development Federal Highway Administration 6300 Georgetown Pike McLean, VA 22101-2296		13. Type of Report and Period Covered Laboratory Report, November 2009–May 2012
		14. Sponsoring Agency Code
15. Supplementary Notes The Contracting Officer's Technical Representative (COTR) was Kornel Kerenyi (HRDI-50).		

16. Abstract

Pier scour design in the United States is currently accomplished through application of the Colorado State University (CSU) equation. Since the Federal Highway Administration recommended the CSU equation in 2001, substantial advances have been made in the understanding of pier scour processes. This report explains a new formulation for describing scour processes and proposes a new equation for pier scour design. A critical review of selected studies is summarized. A simplified scour mechanism is proposed in terms of a pressure gradient resulting from the flow-structure, flow-sediment, and sediment-structure interactions. An equilibrium scour depth equation is proposed based on this understanding of the scour mechanism and is validated and refined by a combination of laboratory and field data. The proposed equation is primarily applicable to clear-water scour conditions with non-uniform coarse bed materials.

17. Key Words Bridge scour, CSU equation, Hager number, Local scour, Pier scour, Sediment mixtures, Non-uniform bed material, Coarse bed materials		18. Distribution Statement No restrictions. This document is available to the public through the National Technical Information Service; Springfield, VA 22161	
19. Security Classif. (of this report) Unclassified	20. Security Classif. (of this page) Unclassified	21. No. of Pages 62	22. Price

Form DOT F 1700.7 (8-72)　　　　　　　　　　　Reproduction of completed page authorized

SI* (MODERN METRIC) CONVERSION FACTORS

APPROXIMATE CONVERSIONS TO SI UNITS

Symbol	When You Know	Multiply By	To Find	Symbol
LENGTH				
in	inches	25.4	millimeters	mm
ft	feet	0.305	meters	m
yd	yards	0.914	meters	m
mi	miles	1.61	kilometers	km
AREA				
in^2	square inches	645.2	square millimeters	mm^2
ft^2	square feet	0.093	square meters	m^2
yd^2	square yard	0.836	square meters	m^2
ac	acres	0.405	hectares	ha
mi^2	square miles	2.59	square kilometers	km^2
VOLUME				
fl oz	fluid ounces	29.57	milliliters	mL
gal	gallons	3.785	liters	L
ft^3	cubic feet	0.028	cubic meters	m^3
yd^3	cubic yards	0.765	cubic meters	m^3
NOTE: volumes greater than 1000 L shall be shown in m^3				
MASS				
oz	ounces	28.35	grams	g
lb	pounds	0.454	kilograms	kg
T	short tons (2000 lb)	0.907	megagrams (or "metric ton")	Mg (or "t")
TEMPERATURE (exact degrees)				
°F	Fahrenheit	5 (F-32)/9 or (F-32)/1.8	Celsius	°C
ILLUMINATION				
fc	foot-candles	10.76	lux	lx
fl	foot-Lamberts	3.426	candela/m^2	cd/m^2
FORCE and PRESSURE or STRESS				
lbf	poundforce	4.45	newtons	N
lbf/in^2	poundforce per square inch	6.89	kilopascals	kPa

APPROXIMATE CONVERSIONS FROM SI UNITS

Symbol	When You Know	Multiply By	To Find	Symbol
LENGTH				
mm	millimeters	0.039	inches	in
m	meters	3.28	feet	ft
m	meters	1.09	yards	yd
km	kilometers	0.621	miles	mi
AREA				
mm^2	square millimeters	0.0016	square inches	in^2
m^2	square meters	10.764	square feet	ft^2
m^2	square meters	1.195	square yards	yd^2
ha	hectares	2.47	acres	ac
km^2	square kilometers	0.386	square miles	mi^2
VOLUME				
mL	milliliters	0.034	fluid ounces	fl oz
L	liters	0.264	gallons	gal
m^3	cubic meters	35.314	cubic feet	ft^3
m^3	cubic meters	1.307	cubic yards	yd^3
MASS				
g	grams	0.035	ounces	oz
kg	kilograms	2.202	pounds	lb
Mg (or "t")	megagrams (or "metric ton")	1.103	short tons (2000 lb)	T
TEMPERATURE (exact degrees)				
°C	Celsius	1.8C+32	Fahrenheit	°F
ILLUMINATION				
lx	lux	0.0929	foot-candles	fc
cd/m^2	candela/m^2	0.2919	foot-Lamberts	fl
FORCE and PRESSURE or STRESS				
N	newtons	0.225	poundforce	lbf
kPa	kilopascals	0.145	poundforce per square inch	lbf/in^2

*SI is the symbol for the International System of Units. Appropriate rounding should be made to comply with Section 4 of ASTM E380.
(Revised March 2003)

TABLE OF CONTENTS

LIST OF FIGURES

LIST OF TABLES

LIST OF SYMBOLS

B	Channel width, ft
b	Pier diameter, ft
D	Sediment size in general, ft
D/Dt	Material derivative
D_*	Sediment size (no dimension)
D_i	Sediment size where i percent of the sediment is finer by weight, ft
D_j	Size of sediment just under down-flow jet, ft
D_{50}	Median grain size, ft
F	Froude number (no dimension)
F	Hydrodynamic force, lb
f, f_i	Functional symbol $i = 1, 2, 3$ (no dimension)
F_f	Friction
g	Gravitational acceleration, ft/s^2
H	Hager number defined as $V / \sqrt{(\rho_s / \rho - 1) g D_{50}}$ (no dimension)
H_c	Critical particle value of H corresponding to the Shields diagram (no dimension)
H_{cp}	Critical value of H for pier scour (no dimension)
h	Flow depth, ft
K_1	Correction for pier shape
K_2	Correction for attack angle of approach flow
K_3	Correction for bed form
K_4	Correction for armoring
p	Pressure, lb/ft^2
Q	Flow
R_h	Hydraulic radius, ft
R	Radius of pier $R = b/2$, ft
r	Distance from center of pier, ft
t	Time, s
u	Velocity vector, ft/s
u	Velocity distribution, ft/s
u_r	Potential velocity in radial direction, ft/s

u_ϕ	Potential velocity along perimeter, ft/s
v	Kinematic viscosity, ft^2/s
V	Approach flow velocity, ft/s
V_c	Critical approach velocity at sediment threshold, ft/s
V_j	Down-flow jet with velocity, ft/s
W	Submerged weight of sediment, lb
y	Distance from bed, ft
y_s	Scour depth, ft
γ	Specific weight of water, lb/ft^3
ρ	Density of water, slug/ft^3
ρ_s	Density of sediment, slug/ft^3
σ	Sediment non-uniformity (gradation coefficient) defined as $\sqrt{D_{84}/D_{16}}$ (no dimension)
σ_{cp}	Non-uniformity coefficient for H_{cp}.
τ_0	Bed shear stress, lb/ft^2
τ_c	Critical shear at sediment threshold, lb/ft^2
τ_1	Grain bed shear, lb/ft^2
ϕ	Angle (no dimension)
Ω	Vorticity (s^{-1})
∇^2	Laplace operator
∇	Vector differential operator

CHAPTER 1. INTRODUCTION

Hydraulics and scour hazards cause over half of the bridge failures in the United States and have been identified by State bridge authorities as one of the top issues in bridge design and maintenance.[1] Current pier scour design in the United States is mainly based on the Colorado State University (CSU) equation, which is described in *Hydraulic Engineering Circular No. 18* (HEC-18).[2] A recent evaluation of bridge scour research indicated a need to change the current method because substantial advances have been made in understanding pier scour processes.[3] The evaluation compared several methods and considered which methods effectively included the variables now believed to determine pier scour characteristics.

The objective of this report is to describe a new method for estimating pier scour based on an understanding of the flow-structure-sediment interactions and to address the weaknesses in earlier methods. This research focuses on clear-water scour at singular piers in non-cohesive sediment mixtures. The approach was to review previous pier scour methods, formulate an improved model for scour formation, and evaluate the new model using laboratory and field data.

CHAPTER 2. LITERATURE REVIEW

Bridge pier scour has been studied worldwide for more than six decades. An early effort in India examined scour depths at groins and pier noses in a study completed in 1949.[4] However, the engineering profession remains unsatisfied with the currently available tools.[3] Pier scour is influenced by many flow, structure, and sediment factors, particularly the complicated vortices and turbulence structures around piers.

Several comprehensive reviews on the topic are available, including those by Sumer, Ettema et al., Sheppard et al., and Tafarojnoruz et al. (See references 3 and 5–7.) Rather than duplicate previous work, this review focuses on selected pier scour models, including work by Laursen, Richardson and Davis, Melville and Chiew, Oliveto and Hager, Sheppard and Miller, and Sheppard et al. (See references 2, 6, and 8–14.) These works provide a range of approaches that are either frequently applied in the United States and abroad or address pier scour by incorporating critical variables not included in other methods.

LAURSEN'S EQUATION

Given a pier in a flume with different flow and sediment conditions, Laursen observed that equilibrium scour depth mainly increases with approach flow depth.[8] Considering variations in pier diameter and sediment size, Laursen developed the relationship shown in figure 1.[9]

$$\frac{b}{y_s} = 5.5 \left[\frac{\left(\dfrac{y_s}{11.5h} + 1 \right)^{7/6}}{\sqrt{\dfrac{\tau_1}{\tau_c}}} - 1 \right]$$

Figure 1. Equation. Laursen's equation.

Where:

y_s = Scour depth, ft.
h = Flow depth, ft.
b = Pier diameter, ft.
τ_1 = Grain bed shear, lb/ft^2.
τ_c = Critical shear at sediment threshold, lb/ft^2.

In Laursen's equation, scour depends on the characteristics of the flow field (depth and bed shear), the structure (pier diameter), and the sediment (critical shear at sediment threshold). For the maximum potential scour depth, y_s at $\tau_1 = \tau_c$, the equation in figure 1 becomes figure 2.

$$\frac{b}{y_s} = 5.5 \left[\left(\frac{y_s}{11.5h} + 1 \right)^{7/6} - 1 \right]$$

Figure 2. Equation. Laursen's equation for potential maximum scour.

Since the scour depth, y_s, is almost always much less than the flow depth, h, (11.5 ft), the equation in figure 2 is approximated by Ettema, et al., as shown in figure 3.[3]

$$\frac{y_s}{\sqrt{bh}} \approx 1.34$$

Figure 3. Equation. Approximate maximum scour.

This result implies that the square root of the product of pier width and flow depth is the appropriate scaling length for pier scour depth.

CSU EQUATION

The widely used CSU equation, which is described in HEC-18, resulted from a series of studies by Shen et al., Richardson and Davis, and Molinas. (See references 2 and 15–17.) Pier scour is estimated as shown in figure 4.

$$\frac{y_s}{b} = 2K_1K_2K_3K_4\left[\left(\frac{h}{b}\right)^{0.35} F^{0.43}\right]$$

Figure 4. Equation. CSU equation.

Where:

F = Froude number ($F = V/(gh)^{0.5}$).
K_1 = Correction for pier shape ($K_1 = 1$ for circular piers).
K_2 = Correction for attack angle of approach flow ($K_2 = 1$ for direct approach flow).
K_3 = Correction for bed form ($K_3 = 1.1$ for clear-water scour).
K_4 = Correction for armoring ($K_4 = 1$ for sand bed material).

For circular piers under clear-water scour conditions, the equation in figure 4 is rewritten as figure 5.

$$\frac{y_s}{b^{0.65}h^{0.35}} = 2.2K_4F^{0.43}$$

Figure 5. Equation. CSU equation for clear-water scour at circular piers.

In this case, the implied length scaling for scour depth is $b^{0.65}h^{0.35}$. In addition, scour depth increases with Froude number and decreases with increasing armoring. Armoring is represented in K_4 as a function of median grain size (D_{50}), D_{95} sediment size where 95 percent of the sediment is finer by weight, and approach velocity (V).

The equations in figure 4 and figure 5 provide reasonable scour depths for many situations with narrow and intermediate piers.[3] Note that a pier is considered narrow when $b/h < 0.71$ and wide when $b/h > 5.0$. However, the use of the Froude number may not be physically representative of scour mechanisms if h is not very small. This is because the Froude number describes the ratio of inertial to gravitational forces on the fluid, while scour is a phenomenon of the interaction of water and sediment at the bed.

4

MELVILLE-CHIEW EQUATION

Based on experiments at the University of Auckland in Auckland, New Zealand and Nanyang Technological University in Singapore, Melville and Chiew concluded that pier scour can be effectively estimated by considering piers as narrow, intermediate, or wide.[11] The maximum potential scour depth, y_s, is scaled by different lengths, as shown in figure 6.[18]

$$y_s = \begin{cases} b & \text{for narrow piers} \\ \sqrt{bh} & \text{for intermediate piers} \\ h & \text{for wide piers} \end{cases}$$

Figure 6. Equation. Melville-Chiew equation.

Scaling for the intermediate pier scour in the equation in figure 6 is similar to that in the equation in figure 3. Melville and Chiew concluded that clear-water scour, y_s, is almost proportional to V/V_c, with V representing approach flow velocity and V_c representing the critical approach velocity at the sediment movement threshold.[11] They further concluded that scour decreases as the sediment coarseness, D_{50}/b, increases but that scour is independent of sediment size if $D_{50}/b \le 0.02$ (fine sands).

OLIVETO-HAGER EQUATION

Hager and Oliveto developed a scour relationship from studies at the Swiss Federal Institute of Technology in Zurich, Switzerland.[12,13,19] It was improved by Kothyari et al.[20] Hager and Oliveto stated that the flow-sediment interaction at the bed is best described by the densimetric particle Froude number, as shown in figure 7.[19]

$$H = \frac{V}{\sqrt{(\rho_s/\rho - 1)gD_{50}}}$$

Figure 7. Equation. Hager number.

Where:

H = Hager number (densimetric particle Froude number).
D_{50} = Median grain size, ft.
ρ = Water density, slug/ft^3.
ρ_s = Sediment density, slug/ft^3.
g = Gravitational acceleration, ft/s^2.

This equation represents the effect of buoyancy, $(\rho_s/\rho - 1)g$, on the water-sediment interface. For brevity, H is referred to as the Hager number. Oliveto and Hager also developed the relationship shown in figure 8 for a cylindrical pier.[12,13]

$$\frac{y_s(t)}{b^{2/3}h^{1/3}} = 0.068\frac{H^{1.5}}{\sqrt{\sigma}}\ln\frac{t\sqrt{(\rho_s/\rho - 1)gD_{50}}}{b^{2/3}h^{1/3}}$$

Figure 8. Equation. Oliveto and Hager time-based scour.

Where:

$y_s(t)$ = Scour depth at time t, ft.
σ = Sediment non-uniformity (gradation coefficient).
t = Time, s.

This equation acknowledges the potential effect on scour depth by non-uniform sediments as reflected by the sediment non-uniformity parameter, σ. This parameter may also be referred to as a gradation coefficient. It is defined as the square root of the ratio of sediment size where 84 percent of the sediment is finer by weight (D_{84}) to sediment size where 16 percent of the sediment is finer by weight (D_{16}).

The equation in figure 8 does not indicate an equilibrium scour depth as time approaches infinity, but it is expected that for circular piers, equilibrium scour depth has the functional form shown in figure 9.

$$\frac{y_s}{b^{2/3}h^{1/3}} = f\left(\frac{H^{1.5}}{\sqrt{\sigma}}\right)$$

Figure 9. Equation. Form for maximum scour.

The equation in figure 9 shows scour depth scaled to a combination of flow depth and pier width as is true for Laursen's equation and the CSU equation. In addition, the equation suggests that scour depth increases with the Hager number but decreases with sediment non-uniformity.

SHEPPARD-MELVILLE EQUATION

The result of the evaluation of scour equations by Ettema et al. was to recommend that the CSU equation in HEC-18 be replaced with the Sheppard-Melville equation.[3] This equation is an integration of work by Sheppard and Miller and by Melville.[14,17] For clear-water scour, the Sheppard-Melville equation is described as shown in figure 10.

$$\frac{y_s}{b} = 2.5f_1\left(\frac{h}{b}\right)f_2\left(\frac{V}{V_c}\right)f_3\left(\frac{b}{D_{50}}\right)$$

Figure 10. Equation. Sheppard-Melville equation.

V_c may be estimated as described in HEC-18 as shown in figure 11.[2]

$$V_c = 11.17h^{1/6}D_{50}^{1/3}$$

Figure 11. Equation. Critical velocity.

In the relationship in figure 10, the function f_1 represents the flow-structure interaction, the function f_2 represents the flow-sediment interaction, and the function f_3 represents the sediment-structure interaction. For the maximum potential scour depth, the equation in figure 10 is reduced by Ettema et al. to the equation in figure 12.[3]

$$\frac{y_s}{b} = 2.5 \tanh\left[\left(\frac{h}{b}\right)^{0.4}\right]$$

Figure 12. Equation. Potential maximum scour derived from Sheppard-Melville.

When $(h/b)^{0.4}$ is small (for shallow water or wide piers where $h/b \leq 0.3$ ft), the equation is equivalent to figure 13.

$$\frac{y_s}{b^{0.6}h^{0.4}} = 2.5$$

Figure 13. Equation. Scour for shallow water or wide piers.

When $(h/b)^{0.4}$ is large (for deep water or narrow piers $h/b \geq 10$ ft), then the equation is equivalent to figure 14.

$$\frac{y_s}{b} = 2.5$$

Figure 14. Equation. Scour for deep water or narrow piers.

The Sheppard-Melville equation in figure 10 introduces the sediment-structure interaction through $f_3(D_{50}/b)$, but since it is only based on D_{50}, it does not capture the armoring effect for non-uniform bed materials.

SUMMARY

Pier scour results from interactions between flow field, structure, and sediment characteristics. The objective is to capture these effects by the proper selection of independent variables.

The interaction between flow field and structure, as represented by pier width and flow depth, primarily dominates scour depth, as shown in figure 15 where the exponent, λ, is between zero and 1, inclusive.

$$y_s \propto b^\lambda h^{1-\lambda}$$

Figure 15. Equation. Scour proportional to flow and structure parameters.

The interaction between flow field and sediment is a second factor governing scour depth. It is variously represented in terms of the ratio of grain bed shear to critical shear (τ_l/τ_c), the ratio of approach velocity to critical velocity (V/V_c), or a function of the Hager number and sediment non-uniformity (H and σ). Since τ_l/τ_c and V/V_c can be converted to the Hager number through the Shield's diagram and a resistance equation described by Hager and Oliveto, the Hager number with sediment non-uniformity is used to represent the interaction between the flow field and sediment interaction, as shown in figure 16.[19]

$$y_s \propto f(H, \sigma)$$

Figure 16. Equation. Scour proportional to flow and sediment parameters.

The interaction between sediment and structure is a third factor governing scour depth. This factor is typically represented as the ratio of median grain size, D_{50}, to pier width, b. Therefore, considering all three factors, scour depth, y_s, may be described as shown in figure 17.

$$\frac{y_s}{b^\lambda h^{1-\lambda}} = f\left(H, \sigma, \frac{D_{50}}{b}\right)$$

Figure 17. Equation. Scour proportional to sediment and structure parameters.

The specific functional form for the equation in figure 17 is determined based on the understanding of scour mechanisms.

CHAPTER 3. SCOUR MECHANISM

Scour mechanisms have been frequently studied, including by Ettema, Dargahi, Roulund et al., Zhao and Huhe, Dey and Raikar, Unger and Hager, Kirkil et al., and Veerappadevaru et al. (See references 21–30.) Since pressure gradient is responsible for all flow and scour phenomena (including bed shear stress) around piers, this section qualitatively explains scour mechanisms in terms of pressure gradient through the flow-structure interaction, the flow-sediment interaction, and the sediment-structure interaction.

FLOW-STRUCTURE INTERACTION

Flow interacts with a pier in multiple ways. First, a vertical stagnation flow is divided into an up-flow and down-flow jet on the leading face of the pier, as shown in figure 18. Point 1 is the stagnation point between the up-flow and down-flow jets. It is defined as the point of maximum energy from the approach flow at the pier face. Energy at the pier face is the sum of the hydrostatic and kinetic components at any given depth of the approach flow. The down-flow component is directed to the bed. The up-flow component moves toward the level of the approach flow water surface (point 2) and creates a bore wave that increases the water surface elevation at the face of the pier to point 3.

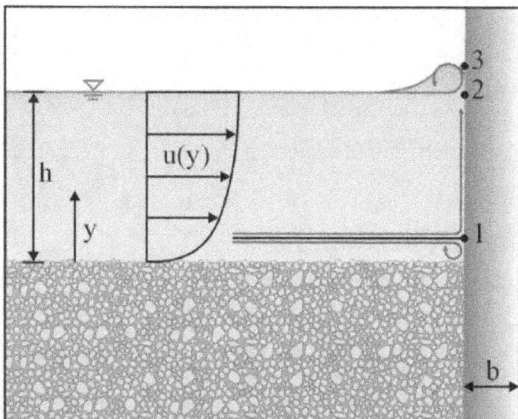

Figure 18. Illustration. Side view of flow-structure interactions in initial scour phase.

In shallow water, the hydrostatic component of the total energy at the pier face is small compared to the kinetic component, resulting in a stagnation point near the water surface and a significant down-flow jet. Scour under this condition is called "shallow water scour." In deep water, the hydrostatic component is greater and, when combined with the kinetic component, tends to create a much more even pressure field at the face of the pier, with the stagnation point closer to the bed and milder up-flow and down-flow jets. Scour under this condition is more weakly related to depth and is called "deep water scour." The scour mechanism treated in this research is intermediate to these extremes.

The second interaction between the structure and the flow field is the creation of two boundary layer flows along the upstream pier perimeter, as shown in figure 19. Assuming the boundary layer flows are fully developed, pressure can be approximated by Bernoulli's equation.

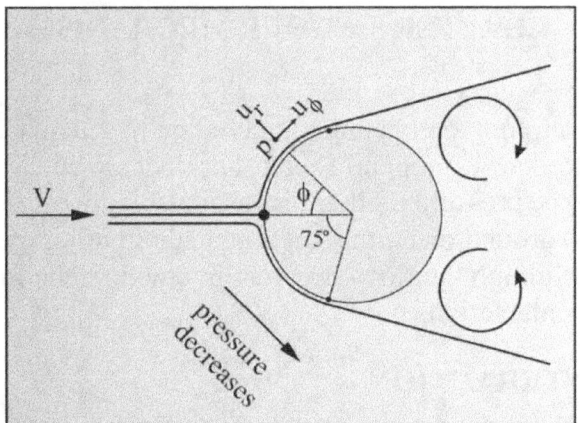

Figure 19. Illustration. Plan view of flow-structure interactions in initial scour phase.

Applying Prandtl's boundary layer theory in conjunction with Bernoulli's equation provides the equation shown in figure 20, as given by Julien.[31]

$$\gamma y + p + \frac{1}{2}\rho\left(u_r^2 + u_\phi^2\right) = \text{constant}$$

Figure 20. Equation. Bernoulli equation with Prandtl boundary layer theory.

Where:

γ = Water specific weight, lb/ft^3.
y = Distance from bed, ft.
p = Pressure, lb/ft^2.
ρ = Density of water, slug/ft^3.
u_r = Potential velocity in the radial direction, ft/s.
u_ϕ = Potential velocity in the tangential direction, ft/s .

The radial and tangential velocities are given in figure 21 and figure 22, respectively.

$$u_r = -u(y)\left(1 - \frac{R^2}{r^2}\right)\cos\phi$$

Figure 21. Equation. Radial velocity.

$$u_\phi = u(y)\left(1 + \frac{R^2}{r^2}\right)\sin\phi$$

Figure 22. Equation. Tangential velocity.

Where:

$u(y)$ = Approach velocity at depth, y, ft/s.
ϕ = Angle from the leading edge.
R = Radius of pier, ft.
r = Distance from the pier center ($r > R$), ft.

Inserting the equations in figure 21 and figure 22 into the equation in figure 20 and simplifying results in the equation in figure 23.

$$\gamma y + p + \frac{1}{2}\rho u^2 \left(\frac{R^4}{r^4} - 2\frac{R^2}{r^2} \cos 2\phi + 1 \right) = \text{constant}$$

Figure 23. Equation. Modified Bernoulli equation.

Further modification to solve for pressure gradient in the ϕ-direction results in figure 24.

$$\frac{\partial p}{\partial \phi} = -2\rho u^2 \left(\frac{R^2}{r^2} \right) \sin 2\phi$$

Figure 24. Equation. Pressure gradient.

The result indicates that $\partial p/\partial \phi \leq 0$ for 0 degrees $\leq \phi \leq 90$ degrees, or the flow-structure interaction results in a favorable pressure gradient in the ϕ-direction in the upstream region of the pier. Note that when applying the equation in figure 24 to beds, a slipping velocity or a theoretical bed is applied so that velocity distribution, $u, \neq 0$ ft/s.

The equation in figure 24 is used to describe scour initiation and bed shear stress around piers. First, the pressure gradient is zero at the stagnation point where $\phi = 0$ degrees and $r = R$, meaning that no sediments move downstream at the leading edge at the beginning of scour if the asymmetrical shape of natural sediments is neglected. Second, the maximum pressure gradient occurs at $\phi = \pm 45$ degrees and $r = R$, meaning that scour begins in the upstream flanks of the pier.

This analysis qualitatively agrees with literature data. Hjorth observed that at $\phi = 45$ degrees, the maximum bed shear stress occurs, which corresponds to the maximum pressure gradient if momentum conservation is considered.[32] Dargahi observed that the scouring begins after 25 s at either side of the cylinder at about ±45 degrees.[23] Roulund et al. calculated that at $\phi = 45–70$ degrees, the maximum bed shear stress occurs.[24] Furthermore, Unger and Hager reported that scour starts at $\phi = 75$ degrees, which is within the range described by White for laminar separation points ($\phi = 71–83$ degrees), where pressure suddenly changes to result in the maximum pressure gradient on the flow line.[27,33]

At the separation points, vortex shedding occurs, forming a wake flow region. The wake flow region is filled with vortices governed by the conservation of vorticity, as described by Kundu and shown in figure 25.[34]

$$\frac{D\Omega}{Dt} = (\nabla \bullet \Omega)u + \nu \nabla^2 \Omega$$

Figure 25. Equation. Conservation of vorticity.

Where:

D/Dt = Material derivative.
Ω = Vorticity vector, s^{-1}.

11

∇^2 = Laplace operator.
∇ = Vector differential operator.
u = Velocity vector, ft/s.
v = Kinematic viscosity, ft/s^2.

The first term on the right side represents the rate of change of vorticity due to stretching and tilting of vortex lines (or tubes). The second term represents the rate of change due to diffusion of vorticities. As shown in figure 26, after parent vortices are created by vortex shedding, the velocity gradients in the three directions stretch and tilt the vortex lines so that the fluid particles spin faster (a ballerina effect) downstream and upward. Bernoulli's equation states that pressure decreases as velocity increases, so pressure in the wake region decreases downstream and upward. In other words, favorable pressure gradients form near the downstream of the pier and along a vortex line upward, immediately moving sediments downstream through bed load and suspended load. This is confirmed by Dargahi's observation that the first scour appears in the wake of the cylinder.[23]

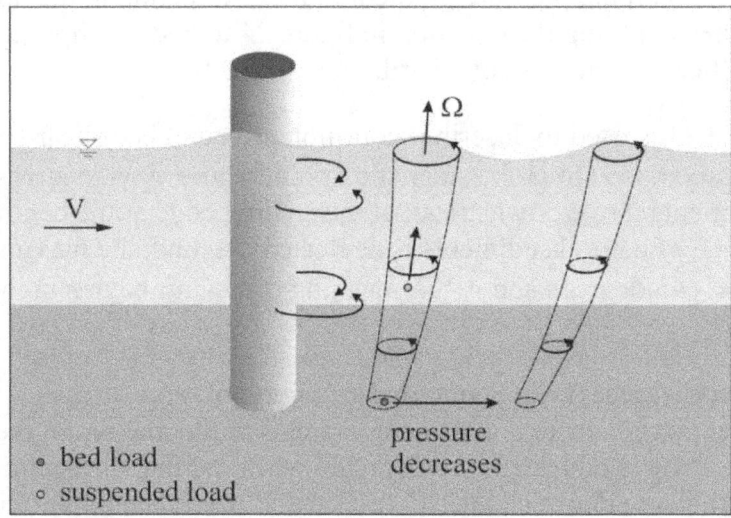

Figure 26. Illustration. Vortex processes in wake flow region.

The vertical stagnation flow is interpreted by the equation in figure 20. Referring to figure 18, Bernoulli's equation between the approach flow ($\phi = 0$, r = ∞) and the pier leading edge ($\phi = 0$, r = R) is shown in figure 27.

$$\gamma y + \gamma (h - y) + \frac{1}{2}\rho u^2 = \gamma y + p$$

Figure 27. Equation. Application of Bernoulli's equation.

This reduces to the equation in figure 28 as follows:

$$p = \gamma (h - y) + \frac{1}{2}\rho u^2$$

Figure 28. Equation. Reduction of Bernoulli equation.

Further simplification is achieved when velocity distribution, u, is approximated by the one-seventh power law, as shown in figure 29.

$$u = \frac{8V}{7}\left(\frac{y}{h}\right)^{1/7}$$

Figure 29. Equation. 1/7th power law.

Substituting the equation in figure 29 into the equation in figure 28 yields a revised relation for the pressure, as shown in figure 30 as follows:

$$p = \gamma\left(h - y\right) + \frac{32\rho V^2}{49}\left(\frac{y}{h}\right)^{2/7}$$

Figure 30. Equation. Stagnation point pressure.

At the stagnation point, the maximum pressure is experienced ($\partial p/\partial y = 0$), and the ratio of the depth in the flow field, y, to the total approach depth, h, is given by a function of the Froude number, F, as seen in figure 31 as follows:

$$\frac{y}{h} = 0.95 F^{2.8}$$

Figure 31. Equation. Ratio for stagnation depth.

If the Froude number equals 0.5, the stagnation point ratio, y/h, is approximately equal to 1 percent, meaning that the stagnation point (point 1 in figure 18) is close to the bed and that only a small part of the approach flow contributes to the down-flow jet in the scour initiation. A large part turns up, forming the up-flow jet and the backwater surface (bore) wave. Near the water surface, the flow moves from point 2 to 3 because pressure at point 2 from the equation in figure 28 is larger than that at point 3 (atmospheric pressure).

FLOW-SEDIMENT INTERACTION

The flow-sediment interaction results from the flow-structure interaction, where pressure gradient determines sediment movement. When the down-flow jet impacts the sediment bed, another stagnation point occurs where the maximum pressure deflects water to upstream of the pier, forming a micro-horseshoe vortex (see figure 18). Unger and Hager observed that such a vortex is usually too weak to initiate scour.[27] Note that this stagnation flow does not return to the perimeter initially because $\partial p/\partial \phi = 0$, as determined from the equation in figure 24.

Referring to figure 32, once scour begins at $\phi = 45-75$ degrees, a series of events is set in motion. The favorable pressure gradient along the perimeter of the pier pushes the adjacent sediment downstream so that scour grows backward to the stagnation point from both sides. Meanwhile, the favorable pressure gradient resulting from the tornado-like vortices pushes wake region sediment downstream through the bed and suspended load, as shown in figure 26, to a low pressure zone where large vortices are broken into small eddies by viscosity resulting in deposition (see figure 32). The decay of the vortices into small eddies is represented by the second term in the equation in figure 25. A scour ring results around the pier to trap the micro-horseshoe vortex that is divided in two along the perimeter. The combination of the horseshoe vortices and the favorable pressure gradient in the ring accelerates scour development. Simultaneously, the combination of the horseshoe vortex and the shedding vortices enhances the scour and transport potential in the wake region.

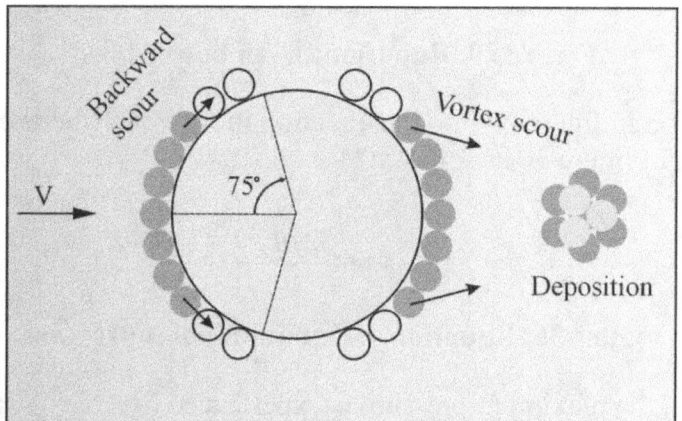

Figure 32. Illustration. Plan view of initial scour phase.

As the scour hole grows, the approach velocity profile is redistributed, as shown in figure 33, so that stagnation point 1 on the pier's leading face shifts up.[27] As a result, the down-flow jet becomes stronger, as do the horseshoe vortices, causing a rapidly (logarithmically) accelerating scour process. (See references 12, 13, and 21–23.) Once the stagnation point on the leading face moves sufficiently close to the water surface, the down-flow jet and horseshoe vortices stabilize. Further scour potential is reduced until an equilibrium scour hole is attained in the form of an inverted frustum, with the upstream side slope approximating the sediment static repose angle and the maximum scour depth at the front face of the pier. Determination of the precise location of the stagnation point requires accurate approach velocity profiles and streamline equations, which are beyond the scope of this qualitative analysis.

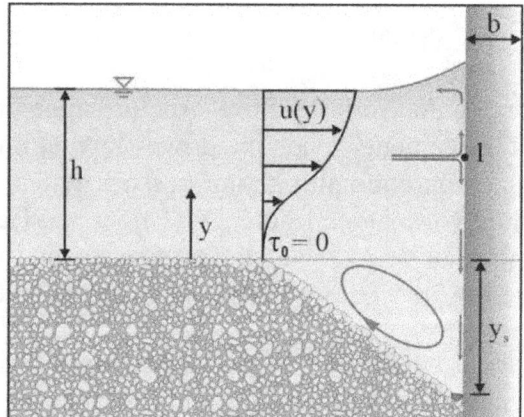

Figure 33. Illustration. Side view of equilibrium phase of scour.

SEDIMENT-STRUCTURE INTERACTION

The sediment-structure interaction is analyzed by considering hydrodynamic forces on individual particles. For qualitative analysis, it is assumed that the feedback of sediment particles on the

14

flow and pressure fields is neglected, and the average pressure difference between the upstream and downstream surface is considered. Figure 34 shows the definition sketch for this analysis.

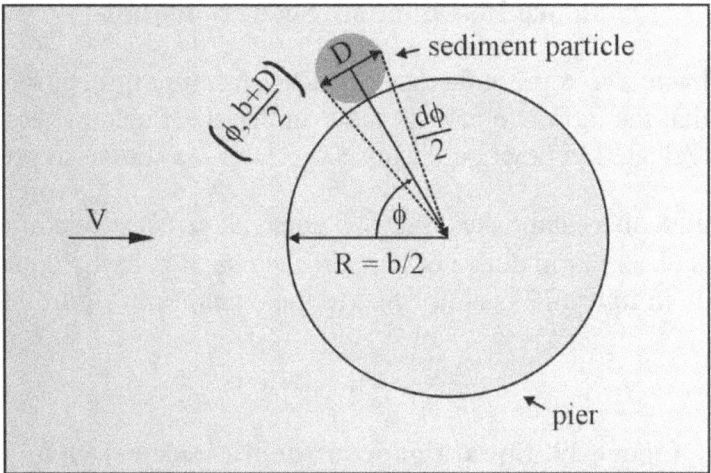

Figure 34. Illustration. Hydrodynamic force of sediment particle.

For a sediment particle located at [ϕ, $(b + D/2)$] (with D representing the sediment size), the pressure gradient at the center of the particle is given by application of the equation in figure 24, resulting in the equation in figure 35.

$$\frac{\partial p}{\partial \phi} = -\frac{2\rho u^2 \sin 2\phi}{(1 + D/b)^2}$$

Figure 35. Equation. Pressure gradient at particle.

This relationship indicates that for a given pier diameter, b, pressure gradient decreases with increasing D. For very large values of D or for particles located far from the pier, no pressure gradient exists, and, therefore, no scour occurs. Approximating $d\phi = D$ [$(b + D/2)$], the pressure difference between the upstream and downstream face is given by the equation in figure 36.

$$\frac{\partial p}{\partial \phi} d\phi = -\frac{4\rho u^2 \sin 2\phi}{(1 + D/b)^3} \frac{D}{b}$$

Figure 36. Equation. Pressure difference.

The hydrodynamic force, F, on a particle is then approximated by the equation in figure 37.

$$F = \left(-\frac{\partial p}{\partial \phi} d\phi\right)\frac{\pi D^2}{4} = \frac{\pi \rho u^2 D^2 \sin 2\phi}{(1 + D/b)^3} \frac{D}{b}$$

Figure 37. Equation. Hydrodynamic force.

The negative sign is used to indicate a favorable pressure gradient for scour. Given D, F increases with b, meaning scour increases with b. Considering that the sediment submerged weight ($W = (\rho_s - \rho)g\pi D^3/6$) resists scour, the local scour potential is described as shown in figure 38.

15

$$\frac{F}{W} = \frac{6\rho u^2}{(\rho_s - \rho)gb} \frac{\sin 2\phi}{(1 + D/b)^3} \propto \frac{V^2}{(\rho_s/\rho - 1)gb} \frac{\sin 2\phi}{(1 + D/b)^3}$$

Figure 38. Equation. Scour potential.

Here, $u \propto V$ was used, and D/b represents the sediment-structure interaction. The equation in figure 38 establishes that the ratio D/b affects scour in the pier flanks where $\phi \neq 0$. However, the effect of D/b can be neglected in practice because $D/b \ll 1$, as shown in previous research.[11]

The equation in figure 38 also establishes that D/b does not affect the scour depth at the leading face where $\phi = 0$. This observation does not support the use of f_3 in the equation in figure 10. Therefore, the equation in figure 17 is simplified to the equation in figure 39.

$$\frac{y_s}{b^\lambda h^{1-\lambda}} = f(\mathrm{H}, \sigma)$$

Figure 39. Equation. Scour depth scale relation.

In summary, the following apply along the perimeter of the pier:

- The flow-structure interaction results in a favorable pressure gradient that pushes sediments to the wake region.

- The vortex processes in the wake region result in a favorable pressure gradient downstream and upward, moving sediments further downstream.

- The down-flow jet from the flow-structure interaction generates horseshoe vortices to scour and move sediments downstream.

- The equilibrium scour depth at the front of the pier, as described by the equation in figure 39, is determined by the flow-structure and the flow-sediment interaction but is independent of the sediment-structure interaction.

EQUILIBRIUM SCOUR DEPTH

The flow-sediment-structure interaction provides a framework for understanding complex three-dimensional (3D) scour processes. If the primary concern is the equilibrium scour depth at the leading edge of the pier, the problem is simplified. As shown in figure 33, the flow collides with the leading face of the pier and generates a down-flow jet. The scour depth, y_s, increases with the area blocking the flow as shown in figure 40, where α represents a positive exponent.

$$y_s \propto (bh)^\alpha$$

Figure 40. Equation. Scour depth and blocking area.

The equation in figure 40 implies that $y_s = 0$ ft if $bh = 0$ ft, which is physically reasonable.

Analyzing equilibrium scour depth as a balancing of forces the down-flow jet with velocity, V_j, applies a lateral hydrodynamic force ($F_l \propto \rho V_j^2 D_j^2$) to the sediment D_j (see figure 41). This force, similar to lift, results from the pressure over the asymmetrical surface of natural sediments and

16

drives the sediment to move, while the submerged sediment weight, W, resists scour through friction, $F_f \propto W$.

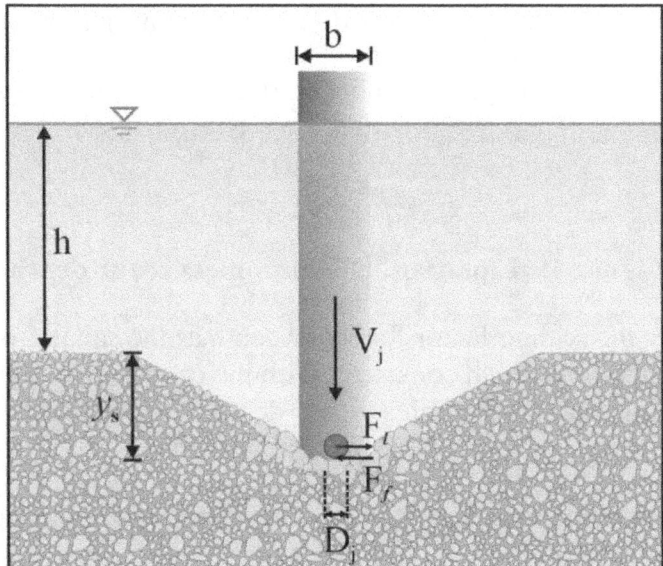

Figure 41. Illustration. Upstream view of equilibrium scour.

Here, $V_j \propto V$ was used. This equation is similar to the equation in figure 38 but results from the down-flow jet.

Experiments reported by Richardson and Davis showed that D_j is approximately D_{90} sediment size where 90 percent of the sediment is finer by weight.[2] Since σ represents $(D_{84}/D_{16})^{0.5}$, σ^2 can be said to equal $(D_{84}/D_{50})(D_{50}/D_{16})$. Assuming a gradation (partially based on the log-normal distribution) where $(D_{84}/D_{50}) = (D_{50}/D_{16}) = (D_{95}/D_{84})$, then $D_{84} = \sigma D_{50}$ and $D_{95} = \sigma^2 D_{50}$. D_{90} may then be approximated as $\sigma^{3/2} D_{50}$. The previous equation then becomes the equation shown in figure 42.

$$\frac{F_l}{F_f} \propto \frac{V^2}{(\rho_s / \rho - 1)gD_{50}\sigma^{3/2}} = \frac{\mathrm{H}^2}{\sigma^{3/2}}$$

Figure 42. Equation. Relative force strength with D_{50}.

This implies the existence of the relationship shown in figure 43.

$$y_s \propto f\left(\frac{\mathrm{H}^2}{\sigma^{3/2}}\right)$$

Figure 43. Equation. Scour depth.

Combining the equations in figure 40 and figure 43 yields the equation in figure 44.

$$y_s = (bh)^\alpha f\left(\frac{H^2}{\sigma^{3/2}}\right)$$

Figure 44. Equation. Scour depth with scaling parameter.

To achieve dimensional homogeneity, the exponent, α, must be 0.5, as shown in figure 45.

$$\frac{y_s}{\sqrt{bh}} = f\left(\frac{H^2}{\sigma^{3/2}}\right)$$

Figure 45. Equation. Dimensionless scour depth.

This relationship defines the scaling factor for scour depth as the square root of the product of pier diameter and approach flow depth consistent with the equations in figure 17 and figure 39, with $\lambda = 0.5$. It also defines the dimensionless scour depth as a function of H and σ, consistent with the equation in figure 9.

The hyperbolic tangent function is selected, as it was for the equation in figure 12, to represent the function shown in figure 46.

$$\frac{y_s}{\sqrt{bh}} = \tanh\left(\frac{H^2/\sigma^{3/2} - H_{cp}^2/\sigma_{cp}^{3/2}}{3.75}\right)$$

Figure 46. Equation. Dimensionless pier scour.

Where:

H_{cp} = Critical Hager number for pier scour.
σ_{cp} = Non-uniformity coefficient for H_{cp}.

Implicit in the development of the critical Hager number is the assumption of uniform sediments so that $\sigma_{cp} = 1$. According to Hager and Oliveto, H_{cp} is determined by the equation in figure 47.[18]

$$H_{cp} = \left[1 - \frac{2}{3}\left(\frac{b}{B}\right)^{1/4}\right] H_c$$

Figure 47. Equation. Critical Hager number for pier scour.

Where:

B = Channel width, ft.
H_c = Critical particle Hager number based on the Shields diagram.

The critical particle Hager number is estimated from figure 48.

$$H_c = \begin{cases} 2.33 D_*^{-0.25} \left(R_h/D_{50}\right)^{1/6} & \text{for } D_* \leq 10 \\ 1.08 D_*^{1/12} \left(R_h/D_{50}\right)^{1/6} & \text{for } 10 < D_* < 150 \\ 1.65 \left(R_h/D_{50}\right)^{1/6} & \text{for } D_* \geq 150 \end{cases}$$

Figure 48. Equation. Critical particle Hager number.

Where:

D_* = Dimensionless sediment size defined as $[(\rho_s/\rho - 1) \cdot g/v^2]^{1/3} \cdot D_{50}$.
R_h = Hydraulic radius, ft.

For a small Hager number (H) or weak a scour, the equation in figure 46 is approximated by the equation in figure 49.

$$\frac{y_s}{\sqrt{bh}} = \frac{H^2/\sigma^{3/2} - H_{cp}^2}{3.75}$$

Figure 49. Equation. Approximation for weak scour.

This relationship suggests that y_s is proportional to H^2 if H_{cp} is neglected. Since H is a function of V, then y_s is proportional to V^2. This agrees with Oliveto and Hager's observation that the scour depth increases with the approach flow velocity, approximately following the square relation.[12]

For a large value of H or strong scour, the equation in figure 46 is approximated by the equation in figure 50.

$$\frac{y_s}{\sqrt{bh}} = 1$$

Figure 50. Equation. Approximation for strong scour.

This relationship corresponds to the maximum potential scour depth and is similar to the equation in figure 3. The relationships developed in this chapter are validated and refined with the data described in chapter 4.

CHAPTER 4. DATA

The development and testing of the concepts in this report are supported by laboratory and field pier scour data. The laboratory data include testing performed specifically for this study at the Federal Highway Administration (FHWA) J. Sterling Jones Hydraulics Laboratory located at the Turner-Fairbank Highway Research Center (TFHRC) as well as data collected at CSU and provided by Molinas.[35] Field data from the United States Geological Survey (USGS) are also used in this study.[36,37]

When considering laboratory and field data jointly, the question of scale is assessed to determine if the same or similar phenomena are captured in the data. In particular, the assessment investigates if the same range of flow, structure, and sediment interaction are captured. Four non-dimensional ratios are summarized in table 1: (1) flow depth to pier width (h/b), (2) pier width to median material size (b/D_{50}), (3) flow depth to median material size (h/D_{50}), and (4) Froude number.

Table 1. Similitude comparison of data sources.

Ratio	Measure	TFHRC	CSU	USGS (2004)	USGS (2011)
h/b	Minimum	1.4	1.0	1.0	0.5
	Median	2.2	1.8	2.1	1.9
	Maximum	6.1	11.2	5.2	5.4
b/D_{50}	Minimum	37	35	11	6
	Median	131	240	13	19
	Maximum	311	393	14	95
h/D_{50}	Minimum	222	353	11	4
	Median	333	435	28	33
	Maximum	444	532	58	272
Froude number	Minimum	0.20	0.10	0.21	0.19
	Median	0.24	0.19	0.47	0.45
	Maximum	0.27	0.28	0.68	1.14

The h/b values are consistent between the four data sources, with the median values narrowly varying between 1.8 in the CSU lab data and 2.2 in the TFHRC lab data. However, the bed material size is much smaller in the two lab data sources (TFHRC and CSU) compared with the two USGS field data sources. Median values of both the b/D_{50} and h/D_{50} ratios are roughly a factor of 10 higher in the field data, meaning the sediment is relatively smaller in the laboratory data. With respect to the flow field, as represented by the Froude number, the two lab datasets exhibit generally lower Froude numbers than the two field datasets.

For this study, the focus is on non-uniform sediments because alluvial sediments are usually composed of a mixture of different sizes of sands and gravels. Previous researchers, including Landers and Mueller, Kranck et al., and Purkait, have proposed that there is a strong tendency for river sediments to follow a log-normal size distribution.[38-40] In contrast, other researchers, including Barndorff-Nielsen, Flenley et al., and Fieller et al., have demonstrated that other distributions, such as the log-skew-Laplace and log-hyperbolic distributions, fit some natural sediment samples better than the log-normal distribution.[41–43] Hajek et al. attributes the differences in the observations of parametric distribution fit to differences in the sediment sampling strategy, measuring, and statistical analysis.[44]

Most natural sediments show an approximate log-normal distribution only through the middle part of the distribution, with long tails in both the coarse and fine fractions. The sediment is characterized by the median grain size D_{50} and a non-uniformity parameter known in literature as the non-uniformity coefficient, σ. This coefficient is defined as the square root of the ratio of D_{84} to D_{16}. The presence of coarse material in sediment mixtures is characterized by D_{90} or D_{95}.

TFHRC LABORATORY DATA

The objective of the experiments was to collect pier scour data under controlled flow conditions for a set of graded and uniform sediment mixtures used as bed material and circular pier models in a flume. The experiments were conducted at the J. Sterling Jones Hydraulics Laboratory located at TFHRC in McLean, VA.

Experimental Setup and Measurements

The facilities, instrumentation, experimental setup, and procedure are described in the following subsections.

Tilting Flume

The experimental flume is a tilting water recirculating laboratory facility. The flume is 6 ft wide, 1.8 ft deep, and 70 ft long, with transparent glass side walls to facilitate flow visualization. It has a stainless steel bottom whose slope was fixed at 0.52 percent for all experiments. A schematic drawing of the flume is presented in figure 51. The numbers in the figure identify the following: (1) pump, (2) magnetic flow meter, (3) stilling basin, (4) honeycomb flow straightener, (5) flow direction, (6) coarse bed material, (7) sediment bed (test section), (8) circular pier structure, and (9) tailgate.

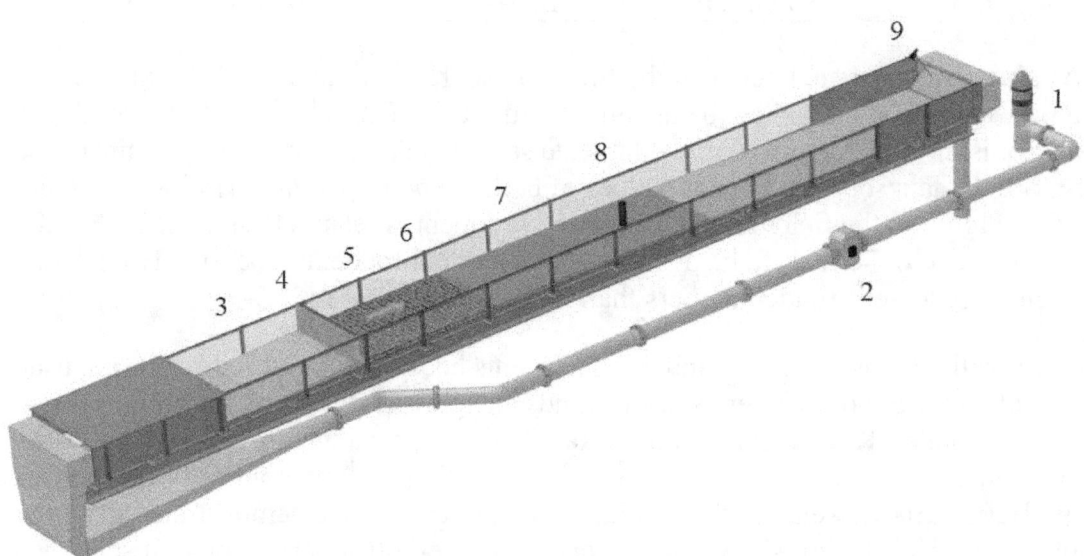

Figure 51. Illustration. FHWA tilting flume.

The structure's skeleton is composed of U-shaped lateral steel frames supported on box-sectioned longitudinal girders. A walkway is provided on one side of the structure. Water is supplied to the flume by a circulation system with a ground sump of 7,420 ft³ and a pump with a maximum capacity of 10.6 ft³/s. The flow is introduced to an upstream head box equipped with a screen and filter. Rapid development of the fully turbulent boundary layer is achieved through an upstream ramp followed by a honeycomb mesh as a flow straightener and an upstream transition zone composed of a layer of coarse sediments carefully placed on the flume bed to provide excess friction. A wave suppressor ensures accomplishment of the previous concerns. The flow depth is regulated through a computer-operated downstream adjustable tailgate.

A uniform 6-inch layer of the test bed material is evenly spread along the full length of the test section (17.4 ft long by 6 ft wide) and starts after the transition zone. A recess where the pier model is positioned is deep enough to model local scour to a depth of 15.7 inches, as shown in figure 52. In order to facilitate drainage of the test section after the experiments, a 1.2-inch-diameter polyvinyl chloride (PVC) pipe was embedded at the bottom of the recess section.

Figure 52. Illustration. Test section with sediment recess at the pier location.

In addition to a manually moving carriage, the flume is instrumented with an automated three-axis positioning system, shown in figure 53, with traversing capability for the entire length, width, and height of the flume and a resolution of 0.0394 inches. This carriage can position probes at any location within the test section to make point measurements for flow velocities using an acoustic Doppler velocimeter (ADV) and bed bathymetry using a laser distance sensor.

Figure 53. Photo. Automated flume carriage in the J. Sterling Jones Hydraulics Laboratory.

Piers

For the pier scour experiments, five cylindrical PVC piers with an outside diameter varying from 0.11 to 0.46 ft were used. For all runs, the piers were placed at the center line of the flume and 23.6 ft from the honeycomb. The largest pier diameter was chosen based on the flume cross section to minimize side-wall effects.

Bed Material

Experimental runs were performed using two types of sediments, a near uniform sediment with D_{50} of 0.018 and 0.035 inches and a mixed sediment with D_{50} of 0.039 and 0.079 inches. The uniform materials were not used for this study. Table 2 provides a summary of the bed material parameters for the mixed sediment materials.

Table 2. Properties of bed materials tested at TFHRC.

Property	Material ID	
	M-1	M-2
Material type	Mixed	Mixed
σ	2.5	2.1
D_{16} (inches)	0.015	0.009
D_{50} (inches)	0.035	0.018
D_{84} (inches)	0.091	0.037
D_{90} (inches)	0.118	0.047
D_{95} (inches)	0.157	0.059
D_{99} (inches) (sediment size where 99 percent of the sediment is finer by weight	0.236	0.091

The relative size of the bed material for comparison with other lab studies and field data may be expressed in two ratios: (1) pier width to D_{50} and (2) flow depth to D_{50}. For the mixed sediment materials, the first ratio ranged from 37 to 311 with a median of 131, and the second ratio ranged from 222 to 444 with a median of 333.

Operating Discharge

To ensure a maximum clear-water pier scour, all approach velocities in the test section were set below the critical velocity, V_c, for D_{50} according to the method proposed by Neill.[45] The upstream velocity was then chosen in the range of $0.93V_c$ to $0.97V_c$ with a constant flow depth, h, equal to 0.66 ft, measured with two ultrasonic sensors placed upstream and downstream along the test section. This resulted in the operating discharge. The discharge was monitored with an electromagnetic flowmeter. An ADV was used for each run to measure point flow velocities at different cross sections along the flume in order to define optimal upstream velocities and validate clear-water flow conditions.

Bed Scour Hole Bathymetry

Bathymetric data were collected for each test bed using a point laser distance sensor shown in figure 54 and figure 55. A LabVIEW™ virtual instrument was programmed for data acquisition and instrument control. First, the initial bed level was mapped right after the sediment bed was leveled prior to the test run. Then, the final bed with the scour hole region was measured after the water was drained and the run was completed. Approximately 2,000 bathymetry points were collected for each pier model for each test. The resulting bathymetry change is calculated as the difference between the initial and final bed elevations.

Figure 54. Photo. Automated point laser distance sensor (side view).

Figure 55. Photo. Automated point laser distance sensor (top view).

Experimental Procedure

Prior to each run, the sediment bed was leveled and compacted with a flat wood plate attached to the manual carriage mounted on the side walls of the tilting flume. The wood plate was the same width as the flume. After the sediment bed was leveled, the pier was carefully placed in the test section. The next step was to map the initial sediment bed bathymetry in the test section around each pier with a point laser distance sensor installed in the automated carriage.

The total size of the mapped bed varied based on the pier diameter with a mesh grid of 0.79 by 0.79 inches. The tailgate was closed, and the flume was slowly filled with water and allowed to sit for approximately 1 h to provide time for air trapped in the sediment to escape. Each test was run for 24 h until equilibrium scour was reached. The pier structure, sediment, and flow conditions for each test are summarized in table 3.

Table 3. Summary of pier scour tests conducted at TFHRC.

Test ID	D_{50} (inches)	Gradation, σ	Pier diameter, b (ft)	Flow, Q (ft^3/s)	Flow Depth, h (ft)	Upstream Velocity, V (ft/s)	Bed Slope (percent)	Froude Number	Flow duration, t (h)
T-1	0.035	2.5	0.46	4.6	0.66	1.25	0.52	0.27	24
T-2	0.035	2.5	0.37	4.6	0.66	1.25	0.52	0.27	24
T-3	0.035	2.5	0.29	4.6	0.66	1.25	0.52	0.27	24
T-4	0.035	2.5	0.20	4.6	0.66	1.25	0.52	0.27	24
T-5	0.035	2.5	0.11	4.6	0.66	1.25	0.52	0.27	24
T-6	0.018	2.1	0.46	3.4	0.66	0.92	0.52	0.20	24
T-7	0.018	2.1	0.37	3.4	0.66	0.92	0.52	0.20	24
T-8	0.018	2.1	0.29	3.4	0.66	0.92	0.52	0.20	24
T-9	0.018	2.1	0.20	3.4	0.66	0.92	0.52	0.20	24
T-10	0.018	2.1	0.11	3.4	0.66	0.92	0.52	0.20	24

Vertical velocity profiles were measured at different cross sections along the test section, and water depths upstream and downstream of the pier were monitored with ultrasonic sensors. At the end of each run, the flume was slowly drained in order to prevent post-run disturbances to the bed and scour hole around the pier. Photos of the scour hole around each pier were taken. A sample result of a test after 24 h is shown in figure 56. The 3D scour hole was mapped using the laser distance sensor, and the collected data were processed with LabVIEW^{TM}. Finally, the dried sediment armored layers formed around the piers were carefully sampled for a sieve analysis using U.S. standard sieves to obtain the particle size distribution of the armor layers.

Figure 56. Photo. Bed bathymetry after a pier scour test.

Experimental Results

The experimental results were based on 3D scour mapping recordings from the sand recess. They are presented in 3D visualizations, longitudinal profiles, and maximum scour depths. The results show that the deepest scour depth was located at the upstream face of the pier, where the strongest down-flow jet occurs. Figure 57 shows the result of a pier scour test with a graded sediment for D_{50} = 0.018 inches, V = 0.92 ft/s, and water depth h = 0.66 ft. Figure 58 shows a close-up photo of the same test showing the armoring layer. A tabulation of experimental results is provided in table 6 in the appendix.

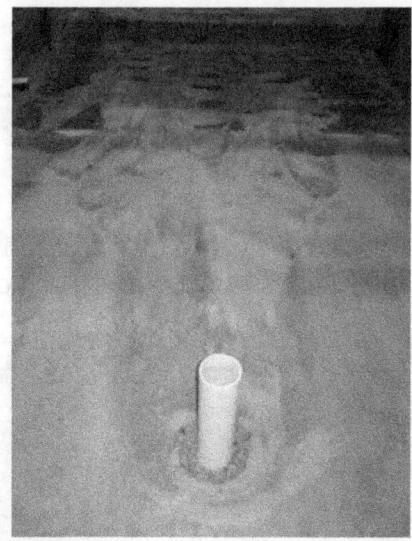

Figure 57. Photo. Result of a pier scour test with a graded sediment.

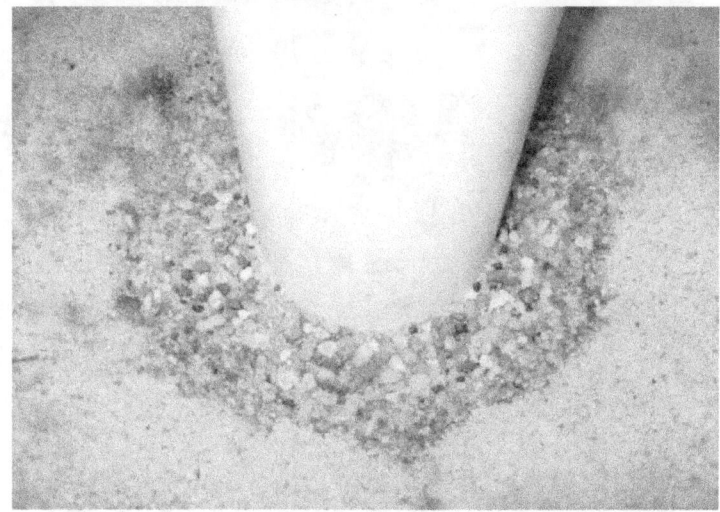

Figure 58. Photo. Result of a pier scour test showing the armoring layer.

CSU DATA

Data collected at CSU by Molinas were used in this study.[35] Table 4 summarizes the properties of the bed material parameters. Table 7 in the appendix summarizes the results of the test runs.

Table 4. Bed material properties for CSU experiments.

Property	Material ID		
	MA-1A	MA-2D	MH-1
σ	2.4	3.4	2.2
D_{16} (inches)	0.012	0.009	0.009
D_{50} (inches)	0.030	0.030	0.022
D_{84} (inches)	0.072	0.104	0.043
D_{90} (inches)	0.083	0.126	0.051
D_{95} (inches)	0.093	0.165	0.063
D_{99} (inches)	0.189	0.315	0.091

In total, 171 tests were conducted, but only 70 were used in this study because they met the following criteria:

- The approach average velocity, V, was less than the critical velocity, V_c, at the sediment D_{50} threshold so that the bed upstream of the pier was not significantly scoured.[2] This ensured that the approach flow conditions did not change during the experiment and the scour was under clear-water conditions. (Given the potential for measurement error, runs with V/V_c less than 1.07 were retained.)

- The flow depth was limited by $1 \leq h/b \leq 6$ to ensure that the scour depth, y_s, was not affected by the bore wave from the up-flow jet and that the vertical stagnation point depended on h.

- Only tests with natural sediment mixtures were used. Uniform bed materials were excluded.

The relative size of the bed material for comparison with other lab studies and field data may be expressed in two ratios: (1) pier width to D_{50} and (2) flow depth to D_{50}. For the CSU bed materials, the first ratio ranged from 35 to 393 with a median of 240, and the second ratio ranged from 353 to 532 with a median of 435.

USGS FIELD DATA

In 2004, Chase and Holnbeck reported field data for pier scour at bridges in Montana, Maryland, and Virginia, and in 2011, Holnbeck reported field data from bridges in Montana.[36,37] Both studies were performed by USGS, and the data are provided in the appendix. All observations are indicative of clear-water scour where approach velocity is less than or equal to critical velocity.

The 2004 bed material ranged in size for D_{50} from 2.2 to 4.3 inches with a gradation ranging from 1.5 to 2.1. This dataset included 22 separate observations; however, it did not include estimates for D_{16}, which is needed to calculate gradation. As an approximation, gradation was estimated as the ratio of D_{95} to D_{50} raised to the reciprocal of 1.645.

As before, the relative size of the bed material for comparison with other lab studies and field data may be expressed in two ratios: (1) pier width to D_{50} and (2) flow depth to D_{50}. For the first ratio, values ranged from 11 to 14 with a median of 13, and the second ratio ranged from 11 to 58 with a median of 28.

The 2011 data from Montana provide a wider range of bed material characteristics than the 2004 data, with D_{50} ranging from 0.4 to 4.3 inches and gradations ranging from 1.5 to 4.1. A total of 89 observations are included in this dataset.

The relative size ratios (pier width to D_{50} and flow depth to D_{50}) also displayed a larger range than the 2004 data. The first ratio ranged from 6 to 95 with a median of 19, and the second ratio ranged from 4 to 272 with a median of 33.

CHAPTER 5. DEVELOPMENT AND ANALYSIS OF DESIGN EQUATION

CONFIRMATION OF GENERAL FORM

The pier scour formulation developed in this study was tested using laboratory and field data. First, a comparison between the various equation forms indicating maximum potential scour and the data is presented. Then, a specific comparison of the data and the proposed scour equation form is provided.

The equation for maximum scour derived from the framework proposed in this study and represented in figure 50 may be compared with equations for maximum scour from Laursen (figure 3), CSU (figure 4 with $K_1 = K_2 = K_4 = 1$, $K_3 = 1.1$, and $F = 0.8$), and Sheppard-Melville (figure 12). The laboratory and field data are shown in figure 59. The figure shows that the proposed framework indicates the lowest normalized scour, y_s/b, for h/b is less than 6, while the Sheppard-Melville equation indicates the lowest normalized scour for h/b is greater than 6. These two relationships provide an envelope for the observed laboratory and field data. The other equations appear to be unnecessarily conservative. The data presented do not contain an observation for normalized scour greater than 2.0. As reported by Ettema et al., 2.5 is the maximum normalized scour.[3]

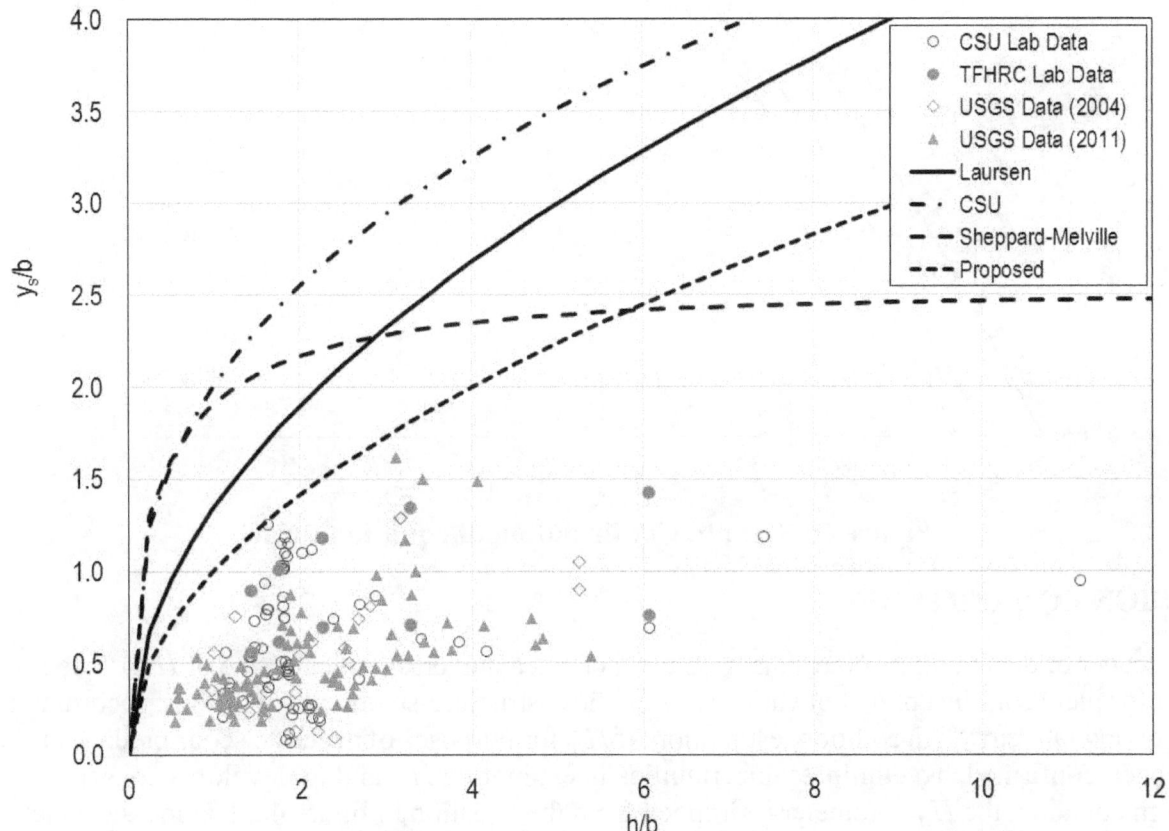

Figure 59. Graph. Comparison of equations for maximum potential scour.

This research suggests that the equation in figure 50 may be used for rough design when h/b is less than 6, and a constant value of 2.5 may be used when h/b is greater than 6 (deep water or narrow piers). This simple framework may also be useful to bridge managers during flood events.

Bridge managers could estimate scour conditions at critical moments and formulate timely corrective strategies. A bridge may be considered at less risk if the estimated scour depth, y_s, is less than the foundation depth. Otherwise, the bridge may be at greater (or critical) risk and should be closed for public safety. Any damage should be immediately repaired after the flood event.

As indicated by the data in figure 59, the equation in figure 50 is likely overly conservative for many design situations. Therefore, further assessment of the more comprehensive equation in figure 46 was conducted. The laboratory and field data are plotted in figure 60 along with the equation in figure 46 using values of H_{cp} equal to 0.0, 0.6, 0.9, and 1.2. Most laboratory data fall into a narrow band between the curves for $H_{cp} = 0.0$ and $H_{cp} = 1.2$. However, much of the field data are above the curves, suggesting that the general form using the hyperbolic tangent formulation is reasonable but could be improved.

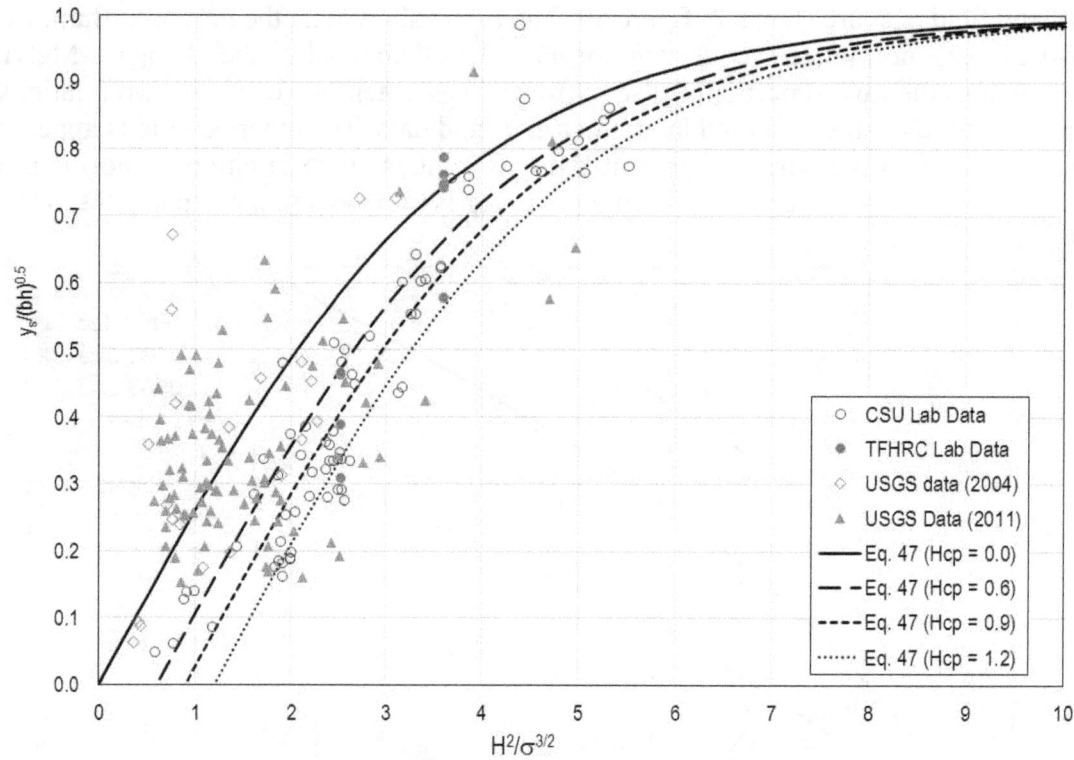

Figure 60. Graph. Confirmation of equation form.

DESIGN EQUATION

The scatter of data in figure 60 results from the fact that a universal critical value of $H^2/\sigma^{3/2}$ does not exist for pier scour inception but varies with the flow-structure-sediment conditions, according to Hager and Oliveto.[18] In addition, estimation of H_{cp} for purposes of accurate scour predictions has not been confirmed. To eliminate uncertainties in estimating H_{cp} and to develop a conservative design equation, the H_{cp} parameter is dropped from the equation in figure 46, leaving the equation in figure 61.

$$\frac{y_s}{\sqrt{bh}} = \tanh\left(\frac{H^2/\sigma^{3/2}}{3.75}\right)$$

Figure 61. Equation. Initial design equation.

Although a conservative design equation is advantageous, it is undesirable to be unnecessarily conservative. To address this, several of the parameters in the equation in figure 61 were optimized considering the laboratory and field data. The equation in figure 62 illustrates the parameters that were optimized.

$$\frac{y_s}{b^\lambda h^{(1-\lambda)}} = A\left[\tanh\left(\frac{H^C/\sigma^D}{B}\right)\right]$$

Figure 62. Equation. Design equation showing optimized parameters.

The parameters for A, B, C, D, and λ were determined as shown in the final design equation in figure 63.

$$\frac{y_s}{b^{0.62}h^{0.38}} = 1.2\left[\tanh\left(\frac{H^2/\sigma^{1.5}}{1.97}\right)\right]$$

Figure 63. Equation. Final design equation.

Using the laboratory data and USGS field data, the relative scour, $y_s/(b^{0.62} \times h^{0.38})$, computed from the final design equation is compared to the measured relative scour depth in figure 64. Of the 190 measured values, 181 were overpredicted, and 9 were underpredicted. Of the overpredicted values, the root mean square (RMS) error was 0.44.

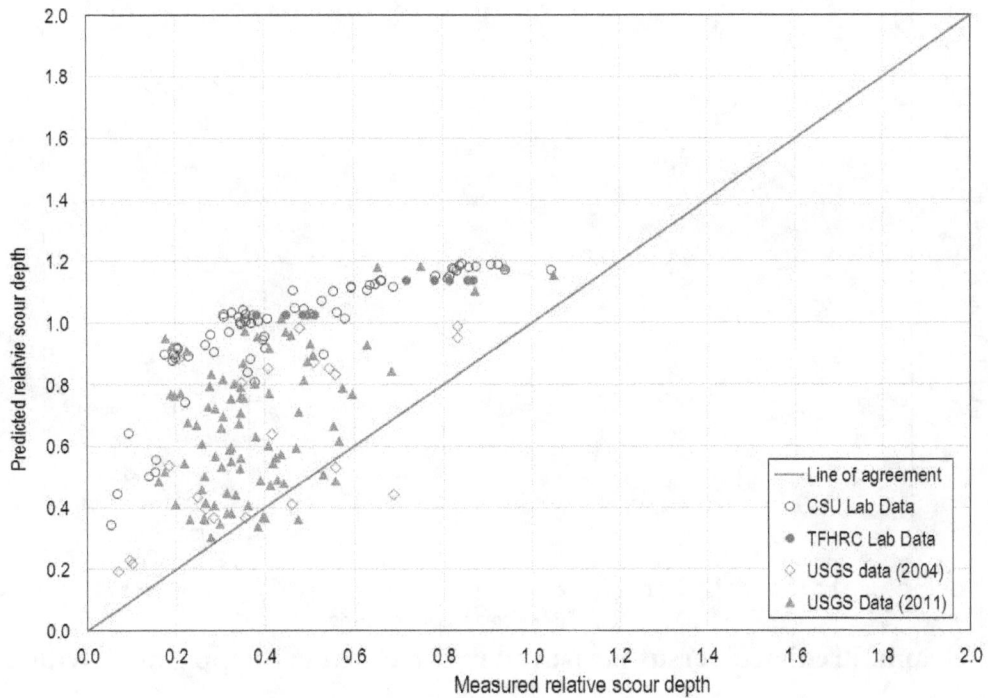

Figure 64. Graph. Predicted versus measured relative scour: proposed equation.

33

The same comparisons were performed using the CSU equation and the Sheppard-Melville equation and are shown in figure 65 and figure 66, respectively. (Critical velocity for the Sheppard-Melville equation was computed using the equation in figure 11.)

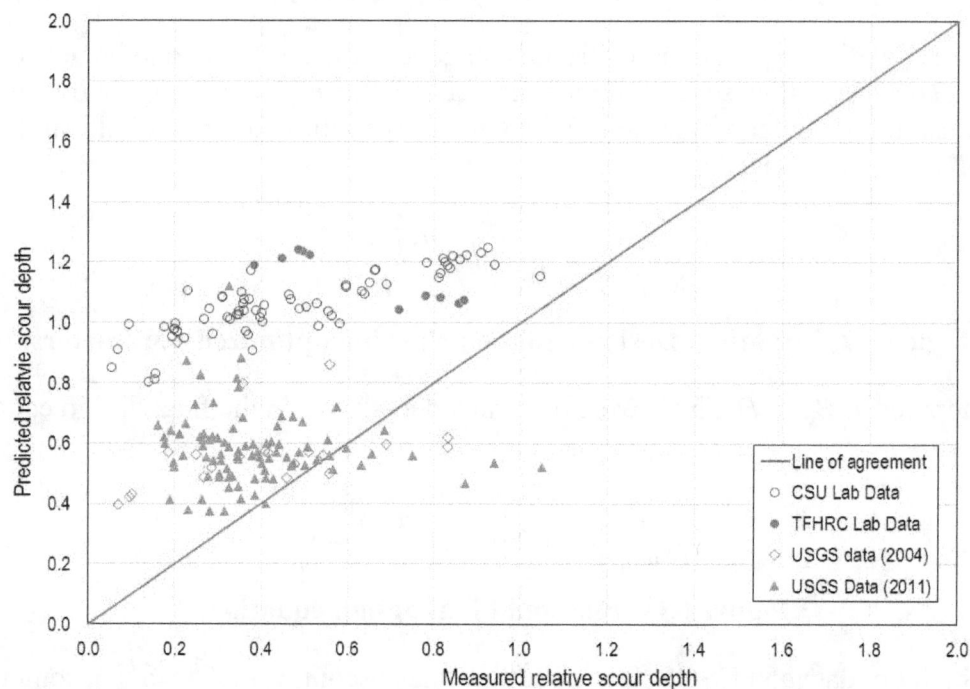

Figure 65. Graph. Predicted versus measured relative scour: CSU equation.

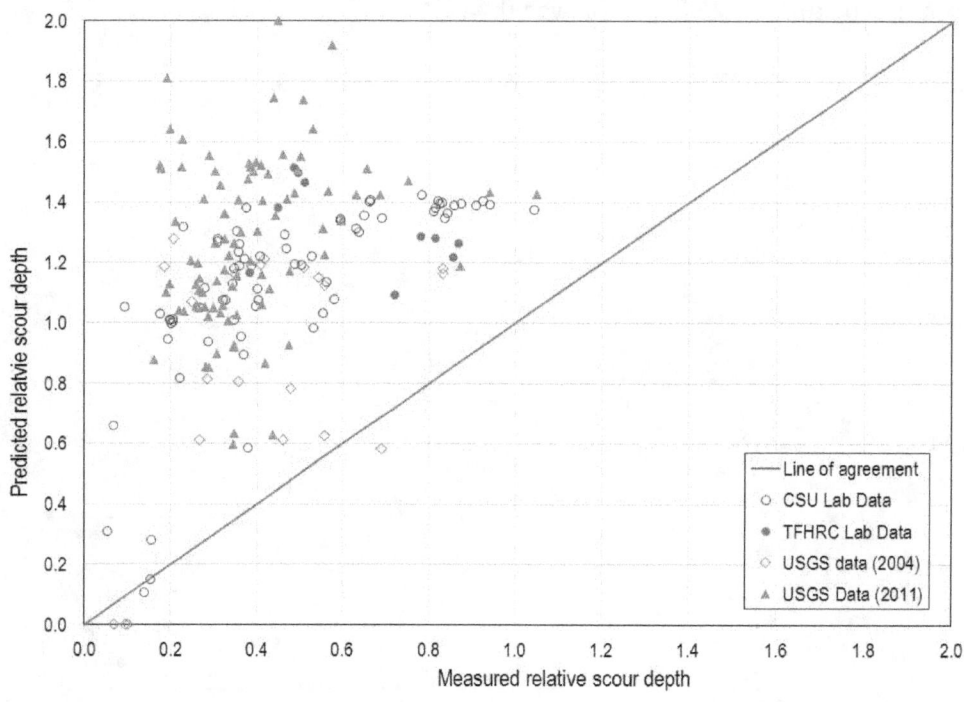

Figure 66. Graph. Predicted versus measured relative scour: Sheppard-Melville equation.

Table 5 provides a summary of the evaluation measures for the proposed design equation and the CSU and Sheppard-Melville equations. The proposed equation is comparable in the number and percent of predictions that are conservative compared with the other two methods. It performs slightly better than the CSU equation and significantly better than the Sheppard-Melville equation in terms of reducing the overprediction, as measured in the RMS error.

Table 5 also provides a reliability index (RI) that measures the risk of underpredicting the actual value. A higher RI indicates a lower risk of underprediction. It is calculated as $(RI) = (1-M_x)/S_x$ where M_x is the mean and S_x is the standard deviation of a series of x values. In this case, x is defined as the ratio of the measured scour to the predicted scour. The risk of underpredicting is lowest with the Sheppard-Melville equation and greatest with the CSU equation.

Compared with the CSU equation, the proposed equation has a lower RMS error and higher RI, indicating that the proposed equation would be an improvement over the CSU equation. Therefore, the proposed equation, formulated based on the sediment-structure-flow interactions, is recommended for use.

Table 5. Comparison of design equations.

Equation	Number Conservative	Percent Conservative	RMS Error for Conservative Predictions	RI
Proposed	181	95.3	0.44	1.85
CSU	176	92.6	0.46	1.33
Sheppard-Melville	184	96.8	0.82	3.23

EQUATION ADJUSTMENTS

The proposed equation in figure 63 is for a simple cylindrical pier. Although no new analyses were conducted as part of this study, the correction factors from HEC-18 for pier nose shape, angle of attack, and bed condition are considered applicable.[2] Therefore, the equation including these factors is presented in figure 67.

$$\frac{y_s}{b^{0.62} h^{0.38}} = 1.2\, K_1\, K_2\, K_3 \left[\tanh \left(\frac{H^2/\sigma^{1.5}}{1.97} \right) \right]$$

Figure 67. Equation. Final design equation with correction factors.

The recommended pier scour design equation was developed by considering scour processes as they are driven by flow, structure, and sediment interactions. It was refined by comparing predicted to measured scour and attempting to minimize the extent of overprediction while remaining conservative, as is appropriate for design.

This equation is most appropriately applied to the following conditions represented in the data used for evaluating the equation:

- Coarse bed materials with 0.39 inches $< D_{50} <$ 4.29 inches.

- Bed material non-uniformity where $1.5 < \sigma < 4.1$.

- Clear-water scour conditions.

Since the equation form was developed from an evaluation of general processes, these ranges should not be considered strict limits but, rather, representative conditions. The designer must consider the applicability of this equation outside these conditions on a site-specific basis.

CHAPTER 6. CONCLUSIONS

After critical review of pier scour under clear-water conditions with non-uniform sediment mixtures, a scour mechanism for understanding pier scour processes and a scour depth equation for design were proposed. In general, pier scour results from flow-structure-sediment interactions, and equilibrium scour depth is determined by the flow-structure and flow-sediment interactions. The following conclusions are presented:

- The flow-structure interaction sets up a favorable pressure gradient along the pier perimeter and in the wake flow region. The perimeter pressure gradient is explained by Prandtl's boundary layer theory and the wake pressure gradient by the motion equation of vorticity. The flow-structure interaction also results in a vertical stagnation flow, generating horseshoe vortices at the foot of the pier and playing an important role in the formation of maximum equilibrium scour depth.

- The flow-sediment interaction results from the flow-structure interaction, where the pressure gradient determines sediment motion. That is, the perimeter pressure gradient and the horseshoe vortices dislodge sediments to the wake of pier, and the wake pressure gradient due to vortex motion further moves sediments downstream to a low pressure zone where large vortices are broken into small eddies.

- The sediment-structure interaction occurs in the flanks of the pier. That is, sediment size affects scour processes but has nothing to do with the equilibrium scour depth at the leading edge.

- Equilibrium scour depth is hypothesized to increase with pier blocking area and Hager number, but it decreases with sediment non-uniformity. This hypothesis was confirmed by the laboratory and field data under various flow-structure-sediment conditions.

- A new clear-water scour design equation was proposed. This equation is conservative, as is appropriate, but it reduces the extent of over-design compared with the CSU and Sheppard-Melville equations and has a higher reliability index than the CSU equation.

- This research is based on a limited database, indicating that extensive experimental research and further analyses are needed for complete understanding of pier scour in clear-water with non-uniform sediment mixtures.

APPENDIX A. DATABASES

This appendix contains tabular summaries of the TFHRC laboratory data (table 6), CSU laboratory data (table 7), and USGS field data (table 8).

Table 6. Summary of TFHRC pier scour tests.

Run ID	Mixture ID	D_{50} (inches)	Gradation, σ	Pier Diameter, b (ft)	Approach Depth, h (ft)	Approach Velocity, V (ft/s)	Measured Scour Depth, y_s (ft)	Flow Duration, t (h)
T-1	M-1	0.035	2.5	0.46	0.66	1.25	0.26	24
T-2	M-1	0.035	2.5	0.37	0.66	1.25	0.23	24
T-3	M-1	0.035	2.5	0.29	0.66	1.25	0.20	24
T-4	M-1	0.035	2.5	0.20	0.66	1.25	0.14	24
T-5	M-1	0.035	2.5	0.11	0.66	1.25	0.08	24
T-6	M-2	0.018	2.1	0.46	0.66	0.92	0.41	24
T-7	M-2	0.018	2.1	0.37	0.66	0.92	0.38	24
T-8	M-2	0.018	2.1	0.29	0.66	0.92	0.34	24
T-9	M-2	0.018	2.1	0.20	0.66	0.92	0.27	24
T-10	M-2	0.018	2.1	0.11	0.66	0.92	0.15	24

Table 7. Summary of CSU laboratory tests.

Run ID	Mixture ID	D_{50} (inches)	Gradation, σ	Pier Diameter, b (ft)	Approach Depth, h (ft)	Approach Velocity, V (ft/s)	Measured Scour Depth, y_s (ft)	Flow Duration, t (h)
MA-1-3	MA-1	0.030	2.4	0.59	1.31	0.66	0.11	8
MA-1-2	MA-1	0.030	2.4	0.59	1.30	0.67	0.12	8
MA-1-1	MA-1	0.030	2.4	0.59	1.26	0.70	0.12	8
MA-4-3	MA-1	0.030	2.4	0.59	1.24	0.95	0.15	12
MA-5-2	MA-1	0.030	2.4	0.59	1.13	0.96	0.15	8
MA-5-1	MA-1	0.030	2.4	0.59	1.16	0.97	0.15	8
MA-5-3	MA-1	0.030	2.4	0.59	1.12	0.97	0.13	8
MA-4-1	MA-1	0.030	2.4	0.59	1.25	0.99	0.16	12
MA-4-2	MA-1	0.030	2.4	0.59	1.23	0.99	0.16	12
MA-6-3	MA-1	0.030	2.4	0.59	1.10	1.08	0.26	12
MA-6-1	MA-1	0.030	2.4	0.59	1.10	1.09	0.29	12
MA-6-2	MA-1	0.030	2.4	0.59	1.10	1.09	0.27	12
MA-9-1	MA-1	0.030	2.4	0.59	1.05	1.10	0.30	16
MA-9-2	MA-1	0.030	2.4	0.59	1.02	1.10	0.26	16
MA-2-3	MA-1	0.030	2.4	0.59	0.96	1.11	0.22	8
MA-9-3	MA-1	0.030	2.4	0.59	1.00	1.12	0.26	16
MA-2-2	MA-1	0.030	2.4	0.59	0.96	1.12	0.22	8
MA-2-1	MA-1	0.030	2.4	0.59	0.94	1.14	0.25	8
MA-7-3	MA-1	0.030	2.4	0.59	1.07	1.27	0.44	16
MA-7-2	MA-1	0.030	2.4	0.59	1.07	1.28	0.44	16
MA-7-1	MA-1	0.030	2.4	0.59	1.06	1.30	0.48	16
MA-8-2	MA-1	0.030	2.4	0.59	1.07	1.35	0.60	12
MA-3-3	MA-1	0.030	2.4	0.59	0.96	1.33	0.47	19
MA-3-2	MA-1	0.030	2.4	0.59	0.95	1.33	0.47	19
MA-8-3	MA-1	0.030	2.4	0.59	1.06	1.38	0.60	12
MA-3-1	MA-1	0.030	2.4	0.59	0.94	1.38	0.55	19
MA-8-1	MA-1	0.030	2.4	0.59	1.05	1.45	0.61	12
MA-10-2	MA-1	0.030	2.4	0.59	1.07	1.50	0.61	10
MA-10-3	MA-1	0.030	2.4	0.59	1.04	1.51	0.60	10
MA-10-1	MA-1	0.030	2.4	0.59	1.09	1.54	0.64	10
MA-12-3	MA-1	0.030	2.4	0.59	1.26	1.58	0.66	16
MA-11-2	MA-1	0.030	2.4	0.59	1.08	1.57	0.65	14
MA-11-1	MA-1	0.030	2.4	0.59	1.10	1.61	0.68	14
MA-12-1	MA-1	0.030	2.4	0.59	1.19	1.65	0.65	16
MA-11-3	MA-1	0.030	2.4	0.59	1.05	1.62	0.68	14
MA-14-1	MA-1A	0.030	2.4	0.59	1.02	0.84	0.16	16
MA-16-1	MA-1A	0.030	2.4	0.59	1.08	0.92	0.27	16
MA-17-1	MA-1A	0.030	2.4	0.59	1.08	0.99	0.30	16
MA-13-1	MA-1A	0.030	2.4	0.59	1.06	1.28	0.51	16
MA-18-1	MA-1A	0.030	2.4	0.59	1.08	1.48	0.70	16

Run ID	Mixture ID	D_{50} (inches)	Gradation, σ	Pier Diameter, b (ft)	Approach Depth, h (ft)	Approach Velocity, V (ft/s)	Measured Scour Depth, y_s (ft)	Flow Duration, t (h)
MA-21-2	MA-2D	0.030	3.4	0.59	1.11	0.69	0.04	16
MA-22-2	MA-2D	0.030	3.4	0.59	1.08	0.79	0.05	16
MA-23-2	MA-2D	0.030	3.4	0.59	1.10	0.98	0.07	16
MA-24-2	MA-2D	0.030	3.4	0.59	1.08	1.24	0.17	16
MA-25-2	MA-2D	0.030	3.4	0.59	1.09	1.43	0.28	16
MH 8-6	MH-1	0.022	2.2	0.06	0.70	0.72	0.06	16
MH 8-5	MH-1	0.022	2.2	0.10	0.64	0.87	0.07	16
MH 7-5	MH-1	0.022	2.2	0.10	0.78	1.00	0.12	16
MH 8-4	MH-1	0.022	2.2	0.19	0.64	0.82	0.12	16
MH 11-4	MH-1	0.022	2.2	0.19	0.72	0.77	0.11	16
MH 7-4	MH-1	0.022	2.2	0.19	0.78	0.91	0.10	16
MH 8-2	MH-1	0.022	2.2	0.29	0.70	0.78	0.22	16
MH 11-2	MH-1	0.022	2.2	0.29	0.78	0.79	0.12	16
MH 6-2	MH-1	0.022	2.2	0.29	0.78	0.91	0.24	16
MH 7-2	MH-1	0.022	2.2	0.29	0.84	0.89	0.25	16
MH 13-1	MH-1	0.022	2.2	0.50	0.69	0.83	0.23	16
MH 16-1	MH-1	0.022	2.2	0.50	0.70	0.93	0.27	16
MH 14-1	MH-1	0.022	2.2	0.50	0.73	0.90	0.29	16
MH 19-1	MH-1	0.022	2.2	0.50	0.73	1.01	0.36	16
MH 12-1	MH-1	0.022	2.2	0.50	0.78	0.92	0.29	16
MH 15-1	MH-1	0.022	2.2	0.50	0.80	1.04	0.38	16
MH 18-1	MH-1	0.022	2.2	0.50	0.81	1.18	0.63	16
MH 8-3	MH-1	0.022	2.2	0.54	0.64	0.87	0.21	16
MH 11-3	MH-1	0.022	2.2	0.54	0.72	0.81	0.16	16
MH 7-3	MH-1	0.022	2.2	0.54	0.78	1.01	0.29	16
MH 8-1	MH-1	0.022	2.2	0.71	0.70	0.84	0.20	16
MH 11-1	MH-1	0.022	2.2	0.71	0.78	0.80	0.15	16
MH 6-1	MH-1	0.022	2.2	0.71	0.78	0.84	0.24	16
MH 5-1	MH-1	0.022	2.2	0.71	0.81	0.95	0.39	16
MH 7-1	MH-1	0.022	2.2	0.71	0.84	0.89	0.26	16

Table 8. Field data measurements.

Mixture/Location	D_{16} (inches)	D_{50} (inches)	D_{84} (inches)	D_{95} (inches)	Gradation, σ	Pier Diameter, b (ft)	Approach Depth, h (ft)	Approach Velocity, V (ft/s)	Measured Scour Depth, y_s (ft)
Montana 1		3.74	8.0	13.0	2.1	3.40	4.80	8.40	0.80
Montana 2		3.74	8.0	13.0	2.1	3.40	3.30	5.10	1.20
Montana 3		3.74	8.0	13.0	2.1	3.40	3.40	6.20	1.90
Montana 4		2.87	5.1	7.5	1.8	3.10	8.70	8.00	2.50
Montana 5		2.87	5.1	7.5	1.8	3.10	8.30	8.20	2.30
Montana 6		2.87	5.1	7.5	1.8	3.10	6.60	4.90	1.90
Montana 7		2.87	5.1	7.5	1.8	3.20	8.20	7.60	1.60
Montana 8		2.87	5.1	7.5	1.8	3.10	7.80	8.00	1.80
Montana 9		2.87	5.1	7.5	1.8	3.20	6.20	4.80	1.10
Montana 10		2.87	5.1	7.5	1.8	3.10	7.40	3.30	0.30
Montana 11		2.87	5.1	7.5	1.8	3.10	6.80	3.60	0.40
Montana 12		2.87	5.1	7.5	1.8	3.10	6.00	3.50	0.40
Montana 13		2.87	5.1	7.5	1.8	3.10	9.80	9.70	4.00
Montana 14		2.87	5.1	7.5	1.8	3.10	9.80	9.10	4.00
Maryland 15		4.33	8.8	13.8	2.0	5.00	7.90	7.70	1.10
Maryland 16		4.33	8.8	13.8	2.0	5.00	6.80	6.80	1.40
Maryland 17		4.33	8.8	13.8	2.0	5.00	9.90	8.60	2.70
Maryland 18		4.33	8.8	13.8	2.0	5.00	8.00	6.20	1.70
Virginia 20		2.17	3.3	4.3	1.5	2.00	2.50	3.70	1.50
Virginia 21		2.17	3.3	4.3	1.5	2.00	10.50	5.50	2.10
Virginia 22		2.17	3.3	4.3	1.5	2.00	10.50	6.40	1.80
Flathead River near Perma	0.65	1.09	2.2	3.2	1.8	3.60	19.40	4.90	1.92
Flathead River near Perma	0.65	1.09	2.2	3.2	1.8	3.50	10.90	4.51	1.90
Clark Fork near Gold Creek	0.34	1.22	3.1	6.3	3.0	5.00	5.58	4.22	2.09
Little Blackfoot River near Avon	0.52	1.15	2.2	3.1	2.0	1.69	2.48	3.62	0.52

Mixture/Location	D_{16} (inches)	D_{50} (inches)	D_{84} (inches)	D_{95} (inches)	Gradation, σ	Pier Diameter, b (ft)	Approach Depth, h (ft)	Approach Velocity, V (ft/s)	Measured Scour Depth, y_s (ft)
Blackfoot River (old bridge) west of Lincoln	0.07	0.38	1.1	1.5	4.0	3.00	7.80	3.03	1.00
Beaverhead River north of Dillon	0.80	1.41	2.3	3.2	1.7	2.67	1.62	3.62	0.98
Jefferson River north of Three Forks	0.35	0.75	1.2	1.9	1.9	3.33	8.47	3.87	0.90
Blackfoot River (old bridge) west of Lincoln	0.07	0.38	1.1	1.5	4.0	3.00	8.60	3.12	1.41
Big Hole River southwest of Twin bridges	0.82	1.52	2.4	3.2	1.7	3.67	5.25	4.58	1.27
Sun River north of Augusta	1.06	2.26	4.4	6.1	2.0	3.50	2.83	4.77	1.17
Big Hole River near Melrose	1.11	2.34	4.1	6.2	1.9	7.00	4.10	5.15	2.23
Yellowstone River near Pray	1.63	3.63	5.2	6.5	1.8	3.00	6.32	6.48	1.67
Yellowstone River south of Livingston	0.95	2.19	3.5	4.6	1.9	10.00	6.00	5.44	3.13
Boulder River near Basin	1.15	2.28	4.4	6.4	2.0	1.69	2.60	4.82	0.32
Yellowstone River near Pray	1.63	3.63	5.2	6.5	1.8	3.00	7.10	6.66	1.20
Madison River south of Cameron	1.18	3.14	5.9	10.0	2.2	2.80	3.01	5.53	0.86
Swan River east of Ferndale	1.37	2.87	4.5	5.7	1.8	3.00	5.00	5.84	0.80
Madison River south of Cameron	1.18	3.14	5.9	10.0	2.2	2.80	3.22	5.64	0.78
Missouri River near Townsend	0.47	1.17	2.5	3.4	2.3	3.00	8.18	4.74	1.43
Blackfoot River (new bridge) west of Lincoln	0.07	0.38	1.1	1.5	4.0	2.00	5.24	3.04	0.76
Clark Fork near Superior	0.79	2.10	4.1	5.7	2.3	5.00	9.76	5.98	2.10
Yaak River near Troy	1.81	4.25	7.4	10.4	2.0	3.50	3.95	6.51	1.05
Clark Fork near Drummond	0.62	1.09	1.7	2.4	1.7	1.67	3.80	4.12	0.70

Mixture/Location	D_{16} (inches)	D_{50} (inches)	D_{84} (inches)	D_{95} (inches)	Gradation, σ	Pier Diameter, b (ft)	Approach Depth, h (ft)	Approach Velocity, V (ft/s)	Measured Scour Depth, y_s (ft)
Little Blackfoot River near Garrison	0.57	2.26	3.9	5.0	2.6	2.00	2.41	4.87	0.60
Jefferson River north of Twin bridges	1.13	2.39	4.5	5.8	2.0	3.00	7.27	5.98	1.35
Flathead River near Perma	0.65	1.09	2.2	3.2	1.8	3.80	17.80	5.42	2.81
Sun River north of Augusta	1.06	2.26	4.4	6.1	2.0	3.50	3.12	5.18	1.38
Clark Fork above Flathead River, near Paradise	0.64	2.16	4.5	6.3	2.7	3.50	10.70	6.34	2.29
Yaak River near Troy	1.81	4.25	7.4	10.4	2.0	3.50	4.50	6.97	1.00
Big Hole River west of Divide	0.93	2.81	6.1	7.5	2.6	4.00	4.54	6.09	1.36
Clark Fork below Flathead River, near Paradise	0.74	1.23	2.0	2.7	1.7	4.33	14.90	5.65	2.66
Big Hole River east of Wise River	1.59	4.02	6.9	10.4	2.1	2.00	7.51	7.50	1.14
Madison River south of Cameron	1.18	3.14	5.9	10.0	2.2	2.80	3.22	6.09	0.79
Yellowstone River near Pray	1.63	3.63	5.2	6.5	1.8	3.00	8.97	7.61	1.40
Boulder River south of Big Timber	1.70	3.02	4.9	6.7	1.7	3.60	8.71	7.17	0.99
Clark Fork near Superior	0.79	2.10	4.1	5.7	2.3	5.00	8.96	6.43	3.54
Clark Fork below Flathead River, near Paradise	0.74	1.23	2.0	2.7	1.7	4.33	17.60	6.03	6.43
Boulder River at I-90, near Cardwell	0.10	0.67	1.7	3.3	4.1	3.00	4.73	4.00	1.37
Sun River north of Augusta	1.06	2.26	4.4	6.1	2.0	3.50	3.60	5.74	1.50
Big Hole River east of Wise River	1.59	4.02	6.9	10.4	2.1	2.00	7.09	7.79	1.12
Big Hole River east of Wise River	1.59	4.02	6.9	10.4	2.1	2.00	7.41	7.89	1.44

Mixture/Location	D_{16} (inches)	D_{50} (inches)	D_{84} (inches)	D_{95} (inches)	Gradation, σ	Pier Diameter, b (ft)	Approach Depth, h (ft)	Approach Velocity, V (ft/s)	Measured Scour Depth, y_s (ft)
Swan River east of Ferndale	1.37	2.87	4.5	5.7	1.8	3.00	7.36	7.09	1.15
Belt Creek 7 mi south of Belt	0.59	1.76	3.5	4.9	2.4	3.50	6.88	5.97	2.13
South Willow Creek near Harrison	0.34	1.51	3.0	4.3	3.0	1.00	1.85	4.57	0.60
Yellowstone River near Pine Creek	0.43	2.15	4.5	6.9	3.2	3.50	10.30	6.85	2.95
Jefferson River north of Twin bridges	1.13	2.39	4.5	5.8	2.0	3.00	7.83	6.80	2.06
Yellowstone River near Pray	1.63	3.63	5.2	6.5	1.8	3.00	9.77	8.13	1.63
Big Hole River west of Divide	0.93	2.81	6.1	7.5	2.6	4.00	4.76	6.65	1.35
Mill Creek near Pray	1.27	2.65	5.1	6.5	2.0	2.00	2.12	5.73	0.53
Gallatin River west of Bozeman	1.79	2.77	4.3	5.6	1.5	3.00	3.61	6.45	0.68
Clark Fork at Missoula	1.12	2.65	4.8	6.6	2.1	3.25	15.40	8.14	1.95
Madison River near Three Forks	0.78	1.17	1.7	2.2	1.5	3.00	2.73	4.67	1.47
North Fork Blackfoot River near Ovando	1.02	1.97	3.7	5.0	1.9	1.69	2.06	5.33	0.45
Mill Creek near Pray	1.27	2.65	5.1	6.5	2.0	2.00	2.52	6.10	0.75
Clark Fork near Alberton	0.71	1.81	3.8	5.5	2.3	3.00	12.40	7.02	2.11
Kootenai River near Libby	1.08	2.28	3.7	5.0	1.8	4.00	6.93	6.91	1.87
Shields River near Wilsall	0.98	1.99	3.5	4.5	1.9	2.50	1.37	5.04	0.45
Little Blackfoot River near Garrison	0.57	2.26	3.9	5.0	2.6	2.00	2.70	5.98	0.75
Missouri River near Townsend	0.47	1.17	2.5	3.4	2.3	3.00	8.47	5.83	1.23
Clark Fork south of Drummond	0.21	1.35	3.4	4.6	4.0	3.00	5.77	5.79	1.53
Big Hole River west of Divide	0.93	2.81	6.1	7.5	2.6	4.00	4.94	7.21	0.76
Boulder River near Cardwell	0.10	0.67	1.7	3.3	4.1	3.00	4.71	4.45	0.71
Little Blackfoot River near Avon	0.52	1.15	2.2	3.1	2.0	1.69	3.18	5.06	1.47

Mixture/Location	D_{16} (inches)	D_{50} (inches)	D_{84} (inches)	D_{95} (inches)	Gradation, σ	Pier Diameter, b (ft)	Approach Depth, h (ft)	Approach Velocity, V (ft/s)	Measured Scour Depth, y_s (ft)
Jefferson River north of Twin bridges	1.13	2.39	4.5	5.8	2.0	3.00	6.02	7.21	2.33
Blackfoot River (new bridge) west of Lincoln	0.07	0.38	1.1	1.5	4.0	2.00	4.76	3.78	0.84
Wisconsin Creek near Sheridan	0.91	1.88	3.1	4.1	1.8	1.02	1.40	5.27	0.40
Smith River at old truss bridge, southeast of Ulm	0.78	1.54	3.4	4.8	2.1	3.53	7.88	6.59	0.85
Clark Fork below Flathead River, near Paradise	0.74	1.23	2.0	2.7	1.7	4.33	13.50	6.73	7.00
Clark Fork at Turah bridge	0.82	1.94	3.4	4.8	2.0	3.00	6.30	6.97	1.94
Jefferson River west of Three Forks	0.23	0.88	2.2	3.5	3.1	3.30	15.90	6.25	2.08
Boulder River at I-90 near Big Timber	1.85	3.33	5.4	7.7	1.7	3.60	2.44	7.22	0.90
Gallatin River near Logan	0.70	1.31	2.3	3.0	1.8	2.80	5.80	6.24	1.70
Yellowstone River near Pine Creek	0.43	2.15	4.5	6.9	3.2	3.50	11.50	8.25	3.05
Yellowstone River at Emigrant	0.57	1.31	2.7	3.7	2.2	4.00	9.69	6.84	2.81
Boulder River at I-90, near Cardwell	0.10	0.67	1.7	3.3	4.1	3.00	5.67	5.02	2.03
Blackfoot River (new bridge) west of Lincoln	0.07	0.38	1.1	1.5	4.0	2.00	5.04	4.08	1.16
South Boulder River near Cardwell	1.15	2.17	4.4	6.1	1.9	1.60	2.24	6.38	0.64
North Fork Blackfoot River west of Lincoln	0.88	1.44	2.4	3.2	1.6	3.00	2.84	5.81	0.56
Clark Fork below Flathead River, near Paradise	0.74	1.23	2.0	2.7	1.7	4.33	14.80	7.39	6.50
Missouri River near Townsend	0.47	1.17	2.5	3.4	2.3	3.00	10.00	6.83	2.99

Mixture/Location	D_{16} (inches)	D_{50} (inches)	D_{84} (inches)	D_{95} (inches)	Gradation, σ	Pier Diameter, b (ft)	Approach Depth, h (ft)	Approach Velocity, V (ft/s)	Measured Scour Depth, y_s (ft)
Belt Creek 2 mi southeast of Belt	0.91	1.56	3.0	4.3	1.8	2.00	2.86	6.11	1.14
Madison River near Three Forks	0.78	1.17	1.7	2.2	1.5	3.00	1.76	5.22	1.10
Clark Fork below Flathead River, near Paradise	0.74	1.23	2.0	2.7	1.7	4.33	12.50	7.38	4.24
Boulder River near McLeod	1.26	2.19	4.3	5.0	1.8	3.00	4.72	7.64	0.80
Clark Fork below Flathead River, near Paradise	0.74	1.23	2.0	2.7	1.7	4.33	13.90	7.59	5.06
Gallatin River west of Bozeman	1.79	2.77	4.3	5.6	1.5	3.00	4.49	8.33	1.25
South Boulder River near Cardwell	1.15	2.17	4.4	6.1	1.9	1.60	0.80	5.78	0.40
Belt Creek 6 mi south of Belt	0.57	1.50	3.0	4.1	2.3	3.50	2.80	6.45	1.85
North Fork Blackfoot River west of Lincoln	0.88	1.44	2.4	3.2	1.6	3.00	3.94	6.77	1.46

Note: Blank cells indicate data are not available.

ACKNOWLEDGMENTS

This research was supported by the FHWA Hydraulics Research and Development Program with Contract No. DTFH61-11-D-00010 through Genex Systems to the University of Nebraska-Lincoln. Roger Kilgore provided technical editing services.

REFERENCES

1. Lwin, M.M. (2010). *Designing and Maintaining Resilient Highway Bridges*, Proceedings of the Sixth US-Taiwan Bridge Engineering Workshop, Seattle, WA.

2. Richardson, E.V. and Davis, S.R. (2001). "Evaluating Scour at Bridges, 4th Edition," *Hydraulic Engineering Circular No. 18*, Publication No. FHWA-NHI-01-001, Federal Highway Administration, Washington, DC.

3. Ettema, R., Constantinescu, G., and Melville, B. (2011). *NCHRP Web-Only Document 175: Evaluation of Bridge Scour Research: Pier Scour Processes and Predictions*, Transportation Research Board, Washington, DC.

4. Inglis, S.C. (1949). *Maximum Depth of Scour at Heads of Guide Bands and Groynes, Pier Noses, and Downstream Bridges—The Behavior and Control of Rivers and Canals*, Indian Waterways Experimental Station, Poona, India.

5. Sumer, B.M. (2007). "Mathematical Modeling of Scour: A Review," *Journal of Hydraulic Research, 45*(6), 723–735.

6. Sheppard, D.M, Demir, H., and Melville, B. (2011). *NCHRP Report 682: Scour at Wide Piers and Long Skewed Piers*, Transportation Research Board, Washington, DC.

7. Tafarojnoruz, A., Gaudio, R., and Dey, S. (2011). "Flow-Altering Countermeasures Against Scour at Bridge Piers: A Review," *Journal of Hydraulic Research, 48*(4), 441–452.

8. Laursen, E.M. (1958). *Scour at Bridge Crossings*, Bulletin Number 8, Iowa Highway Research Board, Ames, IA.

9. Laursen, E.M. (1963). "Analysis of Relief Bridge Scour," *Journal of Hydraulic Engineering, 89*(3), 93–118.

10. Richardson, E.V. and Davis S.R. (1995). "Evaluating Scour at Bridges, 3rd Edition," *Hydraulic Engineering Circular No. 18*, Publication No. FHWA-IP-90-017, Federal Highway Administration, Washington, DC.

11. Melville, B.W. and Chiew, Y.M. (1999). "Time Scale for Local Scour at Bridge Piers," *Journal of Hydraulic Engineering, 125*(1), 59–65.

12. Oliveto, G. and Hager, W.H. (2002). "Temporal Evolution of Clear-Water Pier and Abutment Scour," *Journal of Hydraulic Engineering, 128*(9), 811–820.

13. Oliveto, G. and Hager, W.H. (2005). "Further Results to Time-Dependent Local Scour at Bridge Elements," *Journal of Hydraulic Engineering, 131*(2), 97–105.

14. Sheppard, D.M. and Miller, W. (2006). "Live-Bed Local Pier Scour Experiments," *Journal of Hydraulic Engineering*, *132*(7), 635–642.

15. Shen, H.W., Schneider, V.R., and Karaki, S.S. (1966). *Mechanics of Local Scour*, U.S. Department of Commerce, National Bureau of Standards, Institute for Applied Technology, Washington, DC.

16. Molinas, A. (2001). *Effects of Gradation and Cohesion on Bridge Scour: Synthesis Report*, Report No. FHWA-RD-99-189, Federal Highway Administration, Washington, DC.

17. Melville, B.W. (1997). "Pier and Abutment Scour—An Integrated Approach," *Journal of Hydraulic Engineering*, *123*(2), 125–136.

18. Kandasamy, J.K. and Melville, B.W. (1998). "Maximum Local Scour Depth at Bridge Piers and Abutments," *Journal of Hydraulic Research*, *36*(2), 183–198.

19. Hager, W.H. and Oliveto, G. (2002). "Shields' Entrainment Criterion in Bridge Hydraulics," *Journal of Hydraulic Engineering*, *128*(5), 538–542.

20. Kothyari, U.C., Hager, W.H., and Oliveto, G. (2007). "Generalized Approach for Clear-Water Scour at Bridge Foundation Elements," *Journal of Hydraulic Engineering*, *133*(11), 1229–1240.

21. Ettema, R. (1980). *Scour at Bridge Piers*, Report No. 216, School of Engineering, University of Auckland, Auckland, New Zealand.

22. Dargahi, B. (1989). "The Turbulent Flow Field Around a Circular Cylinder," *Experiments in Fluids*, *8*, 1–12.

23. Dargahi, B. (1990). "Controlling Mechanism of Local Scouring," *Journal of Hydraulic Engineering*, *116*(10), 1197–1214.

24. Roulund, A., Sumer, B.M., Fredsoe, J., and Michelsen, J. (2005). "Numerical and Experimental Investigation of Flow and Scour Around a Circular Pile," *Journal of Fluid Mechanics*, *534*, 351–401.

25. Zhao, W. and Huhe, A. (2006). "Large-Eddy Simulation of Three-Dimensional Turbulent Flow Around a Circular Pipe," *Journal of Hydrodynamics*, Series B, *18*(6), 765–772.

26. Dey, S. and Raikar, R. (2007). "Characteristics of Horseshoe Vortex in Developing Scour Holes at Piers," *Journal of Hydraulic Engineering*, *133*(4), 399–413.

27. Unger, J. and Hager, W.H. (2007). "Down-Flow and Horseshoe Vortex Characteristics of Sediment Embedded Bridge Piers," *Experiments in Fluids*, *42*, 1–49.

28. Kirkil G., Constantinescu, S.G., and Ettema, R. (2008). "Coherent Structures in the Flow Field Around a Circular Cylinder with Scour Hole," *Journal of Hydraulic Engineering*, *134*(5), 572–587.

29. Kirkil G., Constantinescu, S.G., and Ettema, R. (2009). "DES Investigation of Turbulence and Sediment Transport at a Circular Cylinder with Scour Hole," *Journal of Hydraulic Engineering*, *135*(11), 888–901.

30. Veerappadevaru, G., Gangadharaiah, T., and Jagadeesh, T.R. (2011). "Vortex Scouring Process Around Bridge Pier with a Caisson," *Journal of Hydraulic Research*, *49*(3), 378–383.

31. Julien, P.Y. (2010). *Erosion and Sedimentation*, 2nd Edition, Cambridge University Press, Cambridge, UK.

32. Hjorth, P. (1975). *Studies on the Nature of Local Scour*, Bulletin Series A, No. 46, Department of Water Resources Engineering, University of Lund, Sweden.

33. White, F.M. (1991). *Viscous Fluid Flow*, 2nd Edition, McGraw-Hill, New York, NY.

34. Kundu, P.K. (1990). *Fluid Mechanics*, Academic Press, Waltham, MA.

35. Molinas, A. (2003). *Bridge Scour in Nonuniform Sediment Mixtures and in Cohesive Materials: Synthesis Report*, Report No. FHWA-RD-03-083, Federal Highway Administration, McLean, VA.

36. Chase, K.J. and Holnbeck, S.R. (2004). *Evaluation of Pier Scour Equations for Coarse-Bed Streams*, Scientific Investigations Report 2004-5111, U.S. Department of the Interior, U.S. Geological Survey, Reston, VA.

37. Holnbeck, S.R (2011). *Investigation of Pier Scour in Coarse-Bed Streams in Montana, 2001 Through 2007*, Scientific Investigations Report 2011-5107, U.S. Department of the Interior, U.S. Geological Survey, Reston, VA.

38. Landers, M.N. and Mueller, D.S. (1996). *Channel Scour at Bridges in the United States*, Report No. FHWA-RD-95-184, Federal Highway Administration, Washington, DC.

39. Kranck, K., Smith, P.C., and Milligan, T.G. (1996). "Grain-Size Characteristics of Fine-Grained Unflocculated Sediments II: Multi-Round Distributions," *Sedimentology*, *43*, 589–596.

40. Purkait, B. (2002). "Patterns of Grain-Size Distribution in Some Point Bars of the Usri River, India," *Journal of Sedimentary Research*, *72*, 367–375.

41. Barndorff-Nielsen, O. (1977). "Exponentially Decreasing Distributions for the Logarithm of Particle Size," *Proceedings*, Series A, Royal Society of London, *353*, 401–419.

42. Flenley, E., Fieller, N., and Gilbertson, D. (1981). "The Statistical Analysis of Mixed-Grain Size Distributions from Aeolian Sands in the Libyan Pre-Desert Using Log-Skew Laplace Models," *Desert Sediments Ancient and Modern*, Geological Society of London Special Publication, *35*, 271–280.

43. Fieller, N., Flenley, E., and Olbricht, W. (1992). "The Statistics of Particle Size Data," *Royal Statistical Society Journal*, *41*, 127–146.

44. Hajek, A., Huzurbazar, S.V., Mohrig, D., Lynds, R.M., and Heller, P. L. (2010). "Statistical Characterization of Grain Size Distributions in Sandy Fluvial Systems," *Journal of Sedimentary Research*, *80*, 184–192.

45. Neill, C. (1973). *Guide to Bridge Hydraulics*, University of Toronto Press, Toronto, Ontario.

www.ingramcontent.com/pod-product-compliance
Lightning Source LLC
Chambersburg PA
CBHW080646180526
45168CB00008B/3317